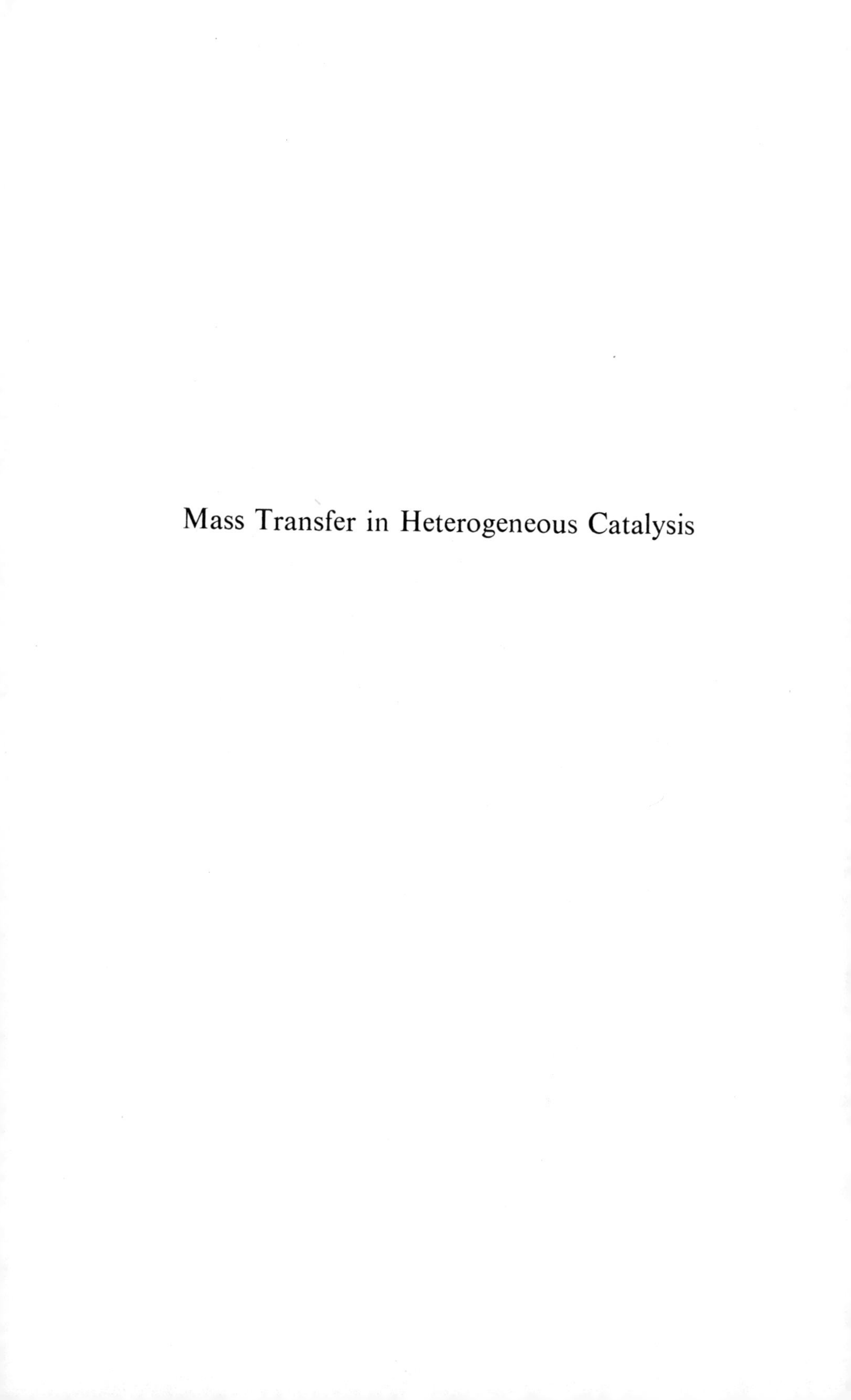

Mass Transfer in Heterogeneous Catalysis

Mass Transfer in Heterogeneous Catalysis

Charles N. Satterfield

ROBERT E. KRIEGER PUBLISHING CO
MALABAR, FLORIDA

Original Edition 1970
Reprint Edition 1981

Printed and Published by
ROBERT E. KRIEGER PUBLISHING COMPANY, INC.
KRIEGER DRIVE
MALABAR, FLORIDA 32950

Printed in the United States of America

Library of Congress Cataloging in Publication Data

Satterfield, Charles N.
Mass Transfer in Heterogeneous Catalysis.

Reprint of the 1970 edition published by M.I.T. Press, Cambridge, Mass.
Bibliography: p.
Includes index.
1. Heterogeneous catalysis. 2. Mass transfer.
I. Title.
[TP156.C35S28 1981] 660.2'995 80-23432
ISBN 0-89874-198-X

To Anne, Mark, and Joye

Contents

List of Examples

List of Tables

Preface

The rate and direction of a chemical reaction are controlled by the manipulation of but a few variables. Beyond the basic ones — pressure, temperature, composition, and contact time — the most common is the use of a catalyst. Catalytic processes have become of such importance in the chemical and petroleum industries that very large sums are being spent, mostly in industrial laboratories, for the development of new chemical syntheses depending on catalysis. The economics of these processes depends greatly on the proper selection and use of the catalyst.

In order for reaction to occur, the reactants must get to the active surface and the products must be removed. This exchange is ordinarily effected by mass transfer or diffusion between the solid catalyst and an ambient stream of gas or liquid. In some cases, the over-all reaction rate is limited by the rate of diffusion, and the likelihood of encountering significant diffusion effects is greatest in the case of the most active catalysts. In this event, the rate of reaction is less than that potentially achievable and, of even greater importance for most applications, the selectivity of the reaction is usually diminished. Optimum reactor design, the selection and use of the catalyst, and the achievement of maximum selectivity require an adequate understanding of the interplay between mass transfer and the intrinsic kinetics of the reaction.

Heterogeneous catalysis usually employs porous solid catalysts, and the stagnant fluid in the pores presents a major diffusional resistance to transport between the ambient stream and the interior of the porous material. The manner in which pore diffusion affects the observed characteristics of a reaction is the subject of an elegant and useful theory, first outlined in this country by Thiele in 1939 and elsewhere by Damköhler and by Zeldovitch. Developed further by Wheeler, and extended by many others, it provides a basic tool to both chemist and design engineer concerned with a catalytic process, and a major portion of this book is devoted to it. Any generalized and reasonably rigorous treatment will be highly complex, but the approach here is to present a simplified model first and then show how this can be modified for various special cases to allow for temperature gradients caused by heat effects, for differences in the mode of diffusion and in the nature of the intrinsic kinetics. The objective has been to develop a clear method of presentation that can be visualized and used readily by the chemist or chemical engineer without the necessity to master involved mathematical techniques. Many results are presented as graphs that also make it possible to relate the reliability of one's final conclusions to the accuracy with which he knows his basic data. Attention is also drawn to the manner in which diffusion can affect the poisoning and regeneration of catalysts as well as the product distribution in the case of competing reactions. A brief discussion is presented of fluidized beds, trickle-bed reactors, and slurry reactors to assist in an understanding of the diffusion effects that may occur in these complex but necessary types of contacting devices.

This book is a major revision and extension of an earlier volume, *The Role of Diffusion in Catalysis*, co-authored with Thomas K. Sherwood and published in 1963. Considerable portions of the present work are contributions from recent research at M.I.T. and the contents of the book represent the outgrowth and refinement of material treated in subjects at M.I.T. over the last several years and in intensive courses and lectures at a variety of industrial organizations.

The comments and suggestions of Professor Sherwood have been a major contribution in the preparation of the present volume. Wayne C. Witkowski and Michael P. Manning have suggested improvements from the viewpoint of the student. The opportunity to write this book was largely made possible by a leave of absence granted by M.I.T. for the academic year 1967–1968.

CHARLES N. SATTERFIELD

Professor of Chemical Engineering
Massachusetts Institute of Technology

Mass Transfer in Heterogeneous Catalysis

1 Diffusion

1.1 Introduction

Most of the important chemical reactions utilized in the chemical and petroleum-refining industries are catalytic and in a major fraction of these the catalyst is a solid substance. The contact process for the manufacture of sulfuric acid, which is now the predominant method, became industrially practicable at the beginning of the twentieth century. World War I saw the first commercial plants for the synthesis of ammonia. Continuous catalytic processes are employed in many large chemical plants, as in the oxidation of ethylene to form ethylene oxide, and napthalene or ortho-xylene to form phthalic anhydride. Styrene is formed by catalytic dehydrogenation of ethyl benzene, butadiene by dehydrogenation of butane or butylene, acrylonitrile by ammoxidation of natural gas. Hydrodesulfurization, cracking, hydrocracking, and reforming are carried out catalytically by the petroleum industry on a very large scale. Continuous processes frequently employ fixed or fluidized beds with reactions in the gas phase, although some hydrogenations and one variation of the Fischer-Tropsch process employ continuous fluidized beds of catalyst suspended in a liquid as a slurry. Trickle beds, in which a liquid flows down through a bed of particles in the presence of a gas phase, are coming into increasing use. The fine-chemical industry employs batch reactors with catalyst suspensions for the hydrogenation of a wide variety of organic chemicals.

Although many millions of dollars are spent each year on the development of new catalysts and new catalytic processes, the phenomena occurring on the catalyst surface are still poorly understood. The selection and manufacture of catalysts for specific purposes is largely empirical, and remains more an art than a science. The investigation of catalytic reactions and catalytic processes is complicated by the fact that diffusion as well as chemical phenomena are usually involved; the two are not easy to separate so as to identify the factors affecting each. One or the other may dominate any particular heterogeneous reaction; the relative importance of these quite different rate processes determines the proper design as well as the performance of catalytic reactors.

In the usual situation, a particle or pellet of solid catalyst is in contact with a gas or liquid in which the reactants are present. The reactants diffuse to the active surface of the catalyst, the reaction takes place, and the products diffuse back to the main body of the ambient fluid. It is usually desirable to employ porous catalysts, since such materials can be made to provide hundreds of square meters of catalytic surface in each gram of solid. If this large internal surface is to be used effectively, the reactants must diffuse first from the fluid to the outside surface of the pellet and then through minute and irregularly shaped pores to the interior.

The pores vary in cross section along their lengths; they branch and interconnect, and some are "dead end." The geometry of the pore structure of common catalysts is poorly defined, and the quantitative treatment of pore diffusion requires a partly empirical approach. Chemical potential decreases in the direction of diffusion through the porous structure, so catalyst surface in the interior of the pellet is in contact with a fluid of lower reactant concentration and higher product concentration than the external or ambient fluid. The internal surface is not as "effective" as it would be if it were all exposed to contact with the external fluid.

Measurements, as of composition and temperature, which the experimenter uses for interpretation of his data are almost invariably those of the bulk of the fluid, yet the observed course of the reaction is the sum of the events occurring throughout the catalyst, being determined by the conditions actually existing at each point on the internal surface of the catalyst. When gradients caused by diffusion are significant, a "falsification of the kinetics" occurs in the sense that the rate and selectivity of the reaction change with bulk concentration and temperature in a different manner than they would in the absence of such gradients. We shall use the term "intrinsic kinetics" to refer to the behavior of the reaction in the absence of diffusional effects and the adjective "apparent" or "effective" to refer to that which is actually observed. When the difference is significant the term "mass transfer limitation" or "diffusion limitation" is frequently used, but this phraseology is subject to misinterpretation. Diffusion causes dissipation of chemical potential, and significant effects of diffusion may be observed when the potential difference between the bulk and the reaction sites amounts to but a few percent of the over-all decrease in potential.

It is difficult to avoid this coupling of physical phenomena with chemical reaction, particularly in industrial processing, where high reaction rates are desired. The scientist or engineer engaged in research or development needs to be able to conduct chemical kinetics studies free of physical transport limitations, if possible, in order to interpret his results correctly, so he must know how to design such experiments properly, be aware of warning signs to look for in his data and have some knowledge of what he may be able to do if diffusion effects become significant. The engineer concerned with development, design, and operation of reactors needs to be aware of what changes in conversion and selectivity may occur as he changes scale or alters operating parameters. Though the interactions between diffusion and chemical reaction are evidently of considerable complexity, many situations are amenable to analysis by the theoretical relations presented in the following chapters. Sufficient experimental data now exist to confirm the value of these theories for application to many practical problems of reactor design.

The general purpose of this book is to describe the role of diffusion in heterogeneous catalysis. The remainder of this chapter is devoted to ordinary molecular or "bulk diffusion," and to diffusion in pores and includes a brief description of typical porous solid catalysts and methods for their physical characterization. Chapter 2 deals with mass transfer between ambient fluid and porous catalyst, for fixed beds, trickle beds, and for particles suspended in liquid or gas. The flow pattern and nature of the contacting is highly complex for most industrial reactors other than one involving a single-phase reactant flowing through a fixed catalyst bed, and even this system is frequently difficult to analyze quantitatively. A brief description is provided in Chapter 2 of fluidized beds, trickle beds, and slurry reactors to assist in the interpretation and prediction of diffusion effects in these systems. The design of packed-bed catalytic reactors, including consideration of radial and axial dispersion, is outside the scope of the present volume. Chapter 3 describes the theory of simultaneous diffusion and reaction in porous structures and the methods for determining the effectiveness factor of a catalyst. The treatment here is based on certain simplifying assumptions; isothermal conditions, a diffusion process following Fick's law, irreversible reaction involving a single reactant and following simple power-law kinetics. The validation of theory by experiment is then discussed. Chapter 4 presents methods of determining the effectiveness factor in more complex situations: those involving temperature gradients inside catalyst pellets, complicated intrinsic kinetic expressions, different catalyst geometries, and volume change upon reaction. Chapter 5 treats the interaction between diffusion and catalyst poisoning and the effects of diffusion on reaction selectivity and on catalyst regeneration. The general approach presented in this book also applies to uncatalyzed heterogeneous reactions such as that of carbon with carbon dioxide or with water vapor, the reduction of iron ore, and to the reaction of solids with acids. The principal differences are that the diffusion process then involves a net flow of mass and the porous character of the solid, if any, changes significantly as reaction proceeds.

1.1.1 *Reaction Regimes*

Visualize a porous solid catalyst pellet in contact with a fluid reactant and consider how the rate of the reaction will change as the temperature is increased. As pointed out by Wicke [384]*, three different catalytic reaction regimes may be observed, as shown diagrammatically on the Arrhenius-type diagram, Figure 1.1, and on Figure 1.2. At sufficiently low temperatures the rate of the reaction will be so low that the potential required to provide the diffusion flux is insignificant and intrinsic kinetics will be observed. With increased temperature, the rate of diffusion per unit potential difference increases but slowly, whereas the intrinsic rate constant increases exponentially. Thus an increasing fraction of the total available potential is required for diffusion, leaving less to drive the chemical reaction. Concentration gradients within the catalyst pores usually become significant before those in the ambient fluid. In this second regime, in which pore diffusion is significant, it will be shown that the apparent activation energy as calculated from an Arrhenius plot will be the arithmetic average of that for the intrinsic reaction and that for diffusion, providing the reaction is simple and the intrinsic kinetics can be expressed by a power-law relationship. In gas phase reactions the effect of temperature on diffusion rates is equivalent to an activation energy of the order of only 1 to 3 kcal, which is so small compared to that of most heterogeneous reactions that the observed activation energy will be little more than one-half the intrinsic value. The apparent order of the reaction will shift toward first order; for example, an intrinsic

* Numbers in brackets are keyed to references at the end of the book.

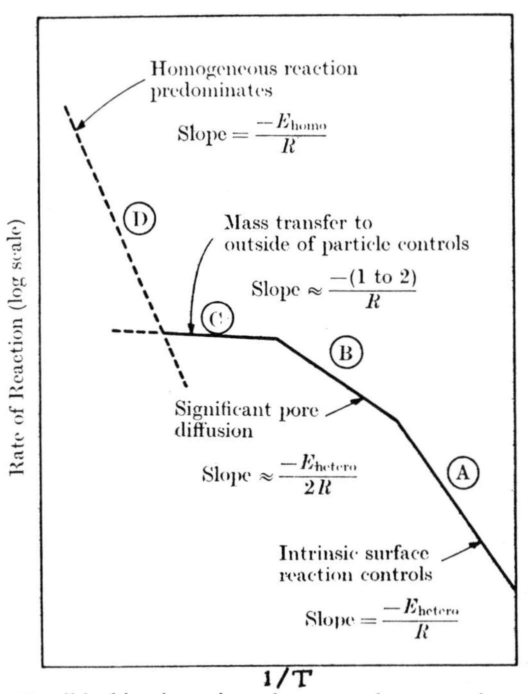

Figure 1.1. Possible kinetic regimes in a gas-phase reaction occurring on a porous solid catalyst.

second-order reaction will appear to be 3/2 order (Section 3.3). Of greatest importance, in complex reactions the selectivity toward an intermediate product will almost invariably be reduced; e.g., in the reaction A → B → C, the yield of B will fall below that otherwise attainable (Section 5.2). The degree of diffusion limitation in this second regime is characterized by the "effectiveness factor" η, defined as the ratio of the observed rate of reaction to that which would occur in the absence of diffusion effects within the pores of the catalyst.

As temperature is further increased, the effectiveness factor becomes progressively smaller, and in another range the concentration difference between the bulk of the fluid and the outside of the catalyst pellet becomes significant. In this third reaction regime the concentration of reactant at the outside surface of the catalyst pellet approaches zero; the rate-limiting process is one of mass transfer from the ambient fluid, and shows the same characteristics as bulk diffusion. The apparent activation energy is then about 1 to 3 for gases, 2.5 to 4.5 in liquid hydrocarbons, and 2.0 to 2.5 in aqueous sytems. In this regime all reactions

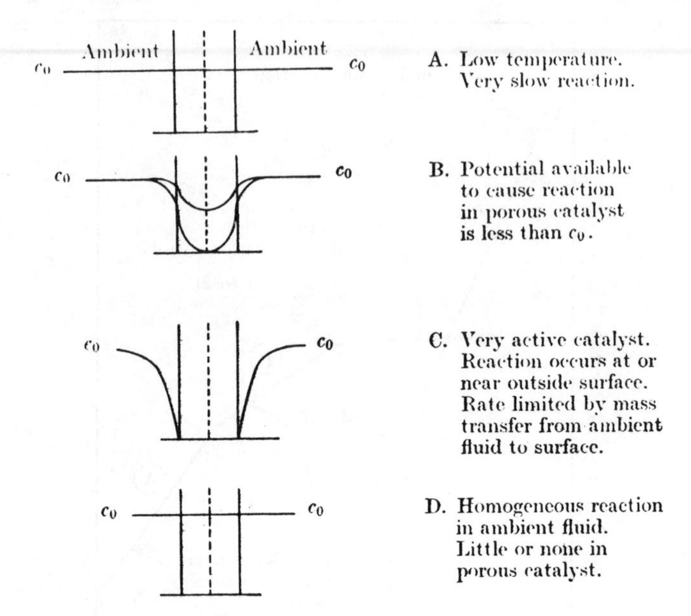

Figure 1.2. Concentration profiles in a porous catalyst under different reaction regimes.

appear to be first order regardless of their intrinsic kinetics, since mass transfer is a first-order process. All catalysts will appear to have the same activity and selectivity, as determined by the relative rates of diffusion of reactants, intermediates, and products rather than by the intrinsic characteristics of the reaction. In this third regime the reaction rate is said to be limited by mass transfer to the catalyst surface, sometimes termed a "film-diffusion-limited" process. This terminology may be confusing. When two processes occur in series, as in regime C, the two rates must be equal under steady-state conditions. The rate-limiting process is the one that consumes the predominant portion of the chemical potential available. In regime B, diffusion through the pore structure occurs simultaneously with reaction. Although reactions in this regime are sometimes described loosely as being "limited by pore diffusion" the process is not controlled by a single step as in regime C.

If the same reaction can occur both homogeneously and catalyzed heterogeneously, the effective activation energy for the homogeneous path is almost invariably greater than that for the heterogeneous route. Since the two competing processes occur in parallel, whichever is inherently faster is the one that will be observed. Homogeneous reaction may predominate over catalytic reaction even at low temperature, depending upon the system, but we show it on the left on Figure 1.1 to emphasize that it will play an increasing role at higher temperatures and that the possibility of contributions from homogeneous reaction must be considered in analyzing the results of a seemingly "catalytic" reaction.

The above picture is somewhat simplified. It represents the intrinsic reaction rate by an Arrhenius expression, whereas more complex kinetics are frequently encountered and indeed, in a few cases, such as the catalytic hydrogenation of ethylene, the intrinsic rate may even exhibit a maximum with increase in temperature. The temperature of the gas and solid are taken to be the same, whereas with high heats of reaction significant temperature gradients between the two may occur, particularly with highly exothermic reactions. Instability effects may develop from the consequences of coupling between temperature and concentration gradients. Nevertheless, the above broad outline describes the transitions from one regime to another as temperature is changed, and has been clearly demonstrated in many experimental studies. (Figure 1.1 represents the separate regimes by intersecting straight lines and omits representation of transition regions.) The order in which the three catalytic regimes will be encountered with increased temperature will be as shown in Figure 1.1, except perhaps in a few highly complex situations. The relative location of the three lines with respect to one another depends on several factors. In the pore-diffusion regime B the rate but not the apparent activation energy will be increased by reducing the size of the catalyst particle or by altering the pore structure so as to increase the diffusivity, provided that this does not move the reaction into another regime: The line will shift upward, its slope remaining the same. With sufficiently small particles, or nonporous solids, this regime will be completely eliminated and transition will occur from intrinsic kinetics directly to regime C. Here the rate is a strong function of the linear velocity of the ambient fluid, whereas in regime B it is independent of it. The relative importance of homogeneous reaction depends upon the ratio of bulk gas volume to catalytic surface as well as the relative rate constants. Broadly useful experimental approaches to determining the regime within which one is operating are thus to observe the effects of particle size, temperature, and agitation or fluid velocity.

The transition temperatures between regimes will vary widely with different reactions and different catalysts. In the burnoff of carbonaceous deposits from a silica-alumina cracking catalyst in the form of 0.20 cm beads, the transition temperature from A to B was 460–475°C (see

Illustrative Example 4.7). The transition region would be at a higher temperature if the particles were smaller. Below this temperature, the rate per unit volume of catalyst did not change as particle size was reduced. An Arrhenius plot of data with the 0.20 cm beads shows the transition expected between regimes A and B. Various other examples of the transition between these two regimes are discussed in Chapter 3. In studies of the combustion of single carbon spheres in air, Hottel and co-workers [339] found that mass transfer to the carbon surface became controlling at temperatures above about 900°C, where, as expected, the observed rate of reaction was strongly affected by gas velocity. An Arrhenius plot showed a pronounced transition between regimes A and C. For the decomposition of nickel carbonyl to deposit metallic nickel on a heated filament, an Arrhenius plot of the rate showed the same type of transition between regimes A and C occurring at about 175°C [65]. In the decomposition of iron pentacarbonyl at 20 Torr (mm Hg) the transition temperature was about 200°C [64]. Deposition of metals by vapor decomposition or other reactions is involved in epitaxial film growth as used in semiconductor technology. In the formation of silicon metal films by reduction of silicon tetrachloride with hydrogen at 1050–1300°C variation in the patterns of deposition on substrates in different furnaces was traced to variation in the natural convection patterns which altered the local rates of this mass-transfer-controlled process. At least three studies have been reported of the dehydrogenation of cyclohexane to benzene on a platinum-on-alumina catalyst [16, 123, 170], in the range of 370–500°C, pressures of about 13 to 42 atm, and with a high ratio of hydrogen to cyclohexane to preserve catalyst activity. At the higher temperatures and pressures the reaction was controlled by mass transfer to the outside of the catalyst particles, regime C [123, 170], but under milder conditions, by diffusion inside catalyst pellets, regime B [16]. A clearly defined transition from intrinsic heterogeneous kinetics with an activation energy of 15 kcal/g-mol to homogeneous reaction with an activation energy of about 55 kcal/g-mol (regime A to regime D) has been shown to occur at about 450°C for the decomposition of hydrogen peroxide vapor at about 7.6 Torr partial pressure (760 Torr total pressure) in contact with Pyrex glass and at a surface-to-volume ratio of 1 cm^{-1} [301]. A marked drop in activation energy upon increasing the temperature of reaction, however, is not of itself proof that a mass transfer limitation is being observed, for such a change may also be brought about solely by a change in chemical mechanism. A good example is the vapor phase oxidation of benzene to maleic anhydride on a V_2O_5-MoO_3/SiO_2 catalyst as reported by

Vaidyanathan and Doraiswamy [340]. The activation energy dropped sharply from 20 kcal/g-mol at temperatures below 350°C to about 2 kcal/g-mol at higher temperatures. Studies with four different catalyst sizes showed that in this case the effect could not be attributed to diffusion limitations. As noted above, the intrinsic catalytic hydrogenation of ethylene and other olefins usually exhibits a maximum rate upon increasing the temperature and then decreases [40]; the same phenomenon has been observed in other hydrogenation reactions.

1.2 Diffusion in Gases

Ordinary molecular diffusion in gases (remote from solid or liquid surfaces) results from differences in concentration between regions of a mixture; diffusion tends to make the concentration uniform throughout.

In a stagnant *binary* gas mixture, the molal flux J (g-mol/sec·cm^2) is proportional to the concentration gradient in the direction of diffusion

$$J_1 = -D_{12}\frac{dc_1}{dx} = -D_{12}c_T\frac{dY_1}{dx}. \tag{1.1}$$

The proportionality constant D_{12} is the diffusion coefficient for gas 1 diffusing in gas 2. This is a function of the molecular properties of the two gases, and increases with increase in temperature or reduction in pressure. The value of D_{12} varies little with the mole fractions Y_1 or Y_2 of the two gases, but does vary with changes in c_T, the total molal concentration of the mixture (g-mol/cm^3). This is Fick's first law.

Motion must be related to some coordinate system. A gas of uniform composition may move as a bulk with respect to fixed coordinates (e.g., an apparatus) so that a flux of the gas can occur without diffusion. However, in many cases net molar transport relative to fixed coordinates occurs even in a stagnant gas, as in the Stefan experiment for measuring diffusivity of a vapor in which a pool of liquid is evaporated at the bottom of a vertical tube and the vapor is allowed to diffuse through a stagnant gas to the top of the tube where it is removed. It is therefore convenient to define a molar flux relative to the plane of *no net molar transport*. The symbol J will be used for this flux, and N for the flux relative to stationary coordinates.

If a binary gas mixture of uniform molal density moves at a constant velocity u_m in the x direction, then Equation 1.1 gives the diffusion flux *relative to the moving gas*:

$$J_1 = (u_1 - u_m)c_1 = -D_{12}c_T\frac{dY_1}{dx}, \tag{1.2}$$

where u_1 is the mean molecular velocity of species 1 relative to fixed coordinates resulting from both bulk motion and diffusion. The flux N_1 is $u_1 c_1$ and the total of the two fluxes (relative to the fixed coordinate system) is $N_1 + N_2$, which is equal to $u_m c_T$. From this, we have

$$N_1 = u_1 c_1 = (N_1 + N_2)Y_1 - D_{12} c_T \frac{dY_1}{dx}. \tag{1.3}$$

A more general result, including a term for the variation of c_1 with time, is

$$\frac{\partial c_1}{\partial t} + u_m \frac{\partial c_1}{\partial x} = D_{12} c_T \frac{\partial^2 Y_1}{\partial x^2} \tag{1.4}$$

or

$$\frac{\partial Y_1}{\partial t} + u_m \frac{\partial Y_1}{\partial x} = D_{12} \frac{\partial^2 Y_1}{\partial x^2}. \tag{1.5}$$

This is for diffusion and flow in the x direction only; the vector form for any direction is

$$\frac{\partial Y_1}{\partial t} + (\boldsymbol{u}_m \cdot \nabla) Y_1 = D_{12} \nabla^2 Y_1. \tag{1.6}$$

In cases where diffusion is accompanied by chemical reaction, the relations must be modified to allow for the formation or disappearance of the diffusing species. Let Q_1 and Q_2 represent the rate of formation of the two species (g-mol/sec$\cdot$cm^3) in a binary system diffusing at constant c_T in the x direction only; then by a molar balance on a differential element,

$$\frac{\partial c_1}{\partial t} + \frac{\partial N_1}{\partial x_1} = Q_1 \tag{1.7}$$

with a similar equation relating c_2, N_2, and Q_2. Adding these equations, and noting that $c_1 + c_2 = c_T$, we obtain

$$Q_1 + Q_2 = \frac{\partial}{\partial x}(N_1 + N_2) = c_T \frac{\partial u_m}{\partial x}. \tag{1.8}$$

Equation 1.3 may be differentiated with respect to x and the derivative used to replace the second term of Equation 1.7; combination with Equation 1.8 then gives

$$\frac{\partial Y_1}{\partial t} + u_m \frac{\partial Y_1}{\partial x} = \frac{Q_1}{c_T} - \frac{Y_1}{c_T}(Q_1 + Q_2) + D_{12} \frac{\partial^2 Y_1}{\partial x^2}, \tag{1.9}$$

where u_m may be replaced by $(N_1 + N_2)/c_T$.

If one mole of species 1 disappears to form one mole of species 2 ($Q_1 = -Q_2$), the second term on the right is absent; if Q_1 and Q_2 are both zero (no reaction), Equation 1.5 results. The general relation in vector form (binary mixture, constant c_T) is

$$\frac{\partial Y_1}{\partial t} + (\boldsymbol{u}_m \cdot \nabla) Y_1 = \frac{Q_1}{c_T} - \frac{Y_1}{c_T}(Q_1 + Q_2) + D_{12} \nabla^2 Y_1. \qquad (1.10)$$

It may be noted that $c_T = P/RT$ in cases where the mixture behaves as an ideal gas.

Equation 1.3 may be integrated for application to several important cases of steady-state diffusion in nonreacting binary gas systems at constant c_T. Thus for the steady-state unidirectional diffusion of species 1 through a stagnant, or nondiffusing, layer of species 2 ($N_2 = 0$),

$$N_1 = -\frac{c_T D_{12}}{1 - Y_1} \frac{dY_1}{dx}. \qquad (1.11)$$

If the gas mixture is ideal, $c_T = P/RT$ and $p_1 + p_2 = P$;

$$N_1 = -\frac{D_{12} P}{RT p_2} \frac{dp_1}{dx} = \frac{D_{12} P}{RT x_0} \ln \frac{(Y_2)_2}{(Y_2)_1}$$

$$= \frac{D_{12} P}{RT x_0} \frac{[(p_1)_1 - (p_1)_2]}{(p_2)_M}. \qquad (1.12)$$

Here x_0 is the thickness of the gas layer, and $(p_2)_M$ is the logarithmic mean of the values of p_2 at $x = 0$ and $x = x_0$, i.e.

$$\frac{(p_2)_2 - (p_2)_1}{\ln [(p_2)_2/(p_2)_1]}.$$

If the two species diffuse at constant but unequal rates, integration of Equation 1.3 gives

$$N_1 + N_2 = \frac{c_T D_{12}}{x_0} \ln \frac{[1 - (Y_1)_2] N_1 - (Y_1)_2 N_2}{[1 - (Y_1)_1] N_1 - (Y_1)_1 N_2}. \qquad (1.13)$$

If the molar diffusion fluxes are constant and equal in opposite directions ($N_1 = -N_2$),

$$N_1 = J_1 = -N_2 = \frac{c_T D_{12}}{x_0} [(Y_1)_1 - (Y_1)_2]. \qquad (1.14)$$

The equations given in the foregoing section apply only to binary gas systems at constant-mixture molar density. The limitation to constant

c_T is not often serious in common application, but many oaooo involving more than two components are encountered. The theory of diffusion in multicomponent mixtures is quite complex, but it has been found possible to deal with the problem in an approximate manner by employing a diffusion coefficient D_{1m} for species 1 in the mixture, related to the binary coefficients by the relation

$$D_{1m} = (1 - Y_1)\left(\sum_{j=2}^{n} \frac{Y_j}{D_{1j}}\right)^{-1}. \tag{1.15}$$

This evidently works well for the diffusion of a single component in a mixture. The analysis of the simultaneous diffusion of two or more components of a mixture is complicated [147]. Where the diffusing components are present in low concentrations, as in many catalytic hydrogenations, the substitution of D_{1m} from Equation 1.15 into equations for single-component diffusion may introduce no great error.

1.3 Diffusion Coefficients: Binary Gas Mixtures

Experimental data on D_{12} for gas systems are available in considerable number; a limited list is given in Table 1.1.

The best present method of extrapolating experimental values, or of estimating coefficients that have not been measured, is the theoretical equation based on the modern kinetic theory and the Lennard-Jones expression for intermolecular forces [147]. This is

$$D_{12} = \frac{0.001858 T^{3/2}[(M_1 + M_2)/M_1 M_2]^{1/2}}{P \sigma_{12}^2 \Omega_D}, \tag{1.16}$$

where T is the absolute temperature (°K), M_1, M_2 are the molecular weights of the two species, P is the total pressure (atm), Ω_D is the "collision integral," a function of $\boldsymbol{k}T/\varepsilon_{12}$ (see Table 1.2), ε, σ are the force constants in the Lennard-Jones potential function, and $\boldsymbol{k}$ is the Boltzmann constant.

Values of ε and σ, usually obtained from viscosity data, are available for a number of pure gases: Table 1.3 presents a selection from the values reported by Svehla [331] for over 200 compounds. Earlier sets of values are reported by Hirschfelder and co-workers [146, 147] and by Rowlinson and Townley [282]. The values of ε and σ for a species should both be taken from the same table. A larger value of ε is substantially compensated for by a smaller value of σ. Where not reported,

Table 1.1 Diffusion Coefficients for Binary Gas Systems
(Experimental values of $D_{12}P$, where D_{12} is in cm^2/sec and P is in atm[a])

Gas Pair	$T(°K)$	$D_{12}P$	Gas Pair	$T(°K)$	$D_{12}P$
Air-ammonia	273	0.198	Ethane-methane	293	0.163
-benzene	298	0.0962	-propane	293	0.0850
-carbon dioxide	273	0.136			
	1000	1.32	Helium-argon	273	0.641
-chlorine	273	0.124	-benzene	298	0.384
-diphenyl	491	0.160	-ethanol	298	0.494
-ethanol	298	0.132	-hydrogen	293	1.64
-iodine	298	0.0834			
-methanol	298	0.162	Hydrogen		
-mercury	614	0.473	-ammonia	298	0.783
-naphthalene	298	0.0611	-benzene	273	0.317
-oxygen	273	0.175	-ethanol	340	0.578
-sulfur dioxide	273	0.122	-ethylene	298	0.602
-toluene	298	0.0844	-methane	288	0.694
-water	298	0.260	-nitrogen	293	0.760
	1273	3.253	-oxygen	273	0.697
			-propane	300	0.450
Argon-neon	293	0.329			
			Nitrogen		
Carbon dioxide-benzene	318	0.0715	-ammonia	298	0.230
-ethanol	273	0.0693	-ethylene	298	0.163
-hydrogen	273	0.550	-iodine	273	0.070
-methane	273	0.153	-oxygen	273	0.181
-methanol	299	0.105	Oxygen-ammonia	293	0.253
-nitrogen	298	0.167	-benzene	296	0.0939
-propane	298	0.0863	-carbon		
			tetrachloride	298	0.071
Carbon monoxide			-ethylene	293	0.182
-ethylene	273	0.151			
-hydrogen	273	0.651	Water-hydrogen	307.2	1.020
-nitrogen	288	0.192	-helium	307	0.902
-oxygen	273	0.185	-methane	307.6	0.292
			-ethylene	307.7	0.204
Dichlorodifluoromethane			-nitrogen	307.5	0.256
-ethanol	298	0.0475	-oxygen	352	0.352
-water	298	0.105	-carbon dioxide	307.4	0.198

[a] A more complete tabulation of published values of D_{12} in gases is to be found in Reference 97.

Table 1.2 Values of the Collision Integral Ω_D Based on the Lennard-Jones Potential [147]

$\frac{kT}{\varepsilon_{12}}$	Ω_D	$\frac{kT}{\varepsilon_{12}}$	Ω_D	$\frac{kT}{\varepsilon_{12}}$	Ω_D
0.30	2.662	1.65	1.153	4.0	0.8836
0.35	2.476	1.70	1.140	4.1	0.8788
0.40	2.318	1.75	1.128	4.2	0.8740
0.45	2.184	1.80	1.116	4.3	0.8694
0.50	2.066	1.85	1.105	4.4	0.8652
0.55	1.966	1.90	1.094	4.5	0.8610
0.60	1.877	1.95	1.084	4.6	0.8568
0.65	1.798	2.00	1.075	4.7	0.8530
0.70	1.729	2.1	1.057	4.8	0.8492
0.75	1.667	2.2	1.041	4.9	0.8456
0.80	1.612	2.3	1.026	5.0	0.8422
0.85	1.562	2.4	1.012	6	0.8124
0.90	1.517	2.5	0.9996	7	0.7896
0.95	1.476	2.6	0.9878	8	0.7712
1.00	1.439	2.7	0.9770	9	0.7556
1.05	1.406	2.8	0.9672	10	0.7424
1.10	1.375	2.9	0.9576	20	0.6640
1.15	1.346	3.0	0.9490	30	0.6232
1.20	1.320	3.1	0.9406	40	0.5960
1.25	1.296	3.2	0.9328	50	0.5756
1.30	1.273	3.3	0.9256	60	0.5596
1.35	1.253	3.4	0.9186	70	0.5464
1.40	1.233	3.5	0.9120	80	0.5352
1.45	1.215	3.6	0.9058	90	0.5256
1.50	1.198	3.7	0.8998	100	0.5130
1.55	1.182	3.8	0.8942	200	0.4644
1.60	1.167	3.9	0.8888	400	0.4170

Table 1.3 Lennard-Jones Force Constants Calculated from Viscosity Data[a]

Compound	ε/k (°K)	σ (Å)
Acetone	560.2	4.600
Acetylene	231.8	4.033
Air	78.6	3.711
Ammonia	558.3	2.900
Argon	93.3	3.542
Benzene	412.3	5.349
Bromine	507.9	4.296

Table 1.3—*Continued*

Compound	ε/k (°K)	σ (Å)
i-Butane	330.1	5.278
Carbon dioxide	195.2	3.941
Carbon disulfide	467	4.483
Carbon monoxide	91.7	3.690
Carbon tetrachloride	322.7	5.947
Carbonyl sulfide	336	4.130
Chlorine	316	4.217
Chloroform	340.2	5.389
Cyanogen	348.6	4.361
Cyclohexane	297.1	6.182
Cyclopropane	248.9	4.807
Ethane	215.7	4.443
Ethanol	362.6	4.530
Ethylene	224.7	4.163
Fluorine	112.6	3.357
Helium	10.22	2.551[b]
n-Hexane	339.3	5.949
Hydrogen	59.7	2.827
Hydrogen cyanide	569.1	3.630
Hydrogen chloride	344.7	3.339
Hydrogen iodide	288.7	4.211
Hydrogen sulfide	301.1	3.623
Iodine	474.2	5.160
Krypton	178.9	3.655
Methane	148.6	3.758
Methanol	481.8	3.626
Methylene chloride	356.3	4.898
Methyl chloride	350	4.182
Mercury	750	2.969
Neon	32.8	2.820
Nitric oxide	116.7	3.492
Nitrogen	71.4	3.798
Nitrous oxide	232.4	3.828
Oxygen	106.7	3.467
n-Pentane	341.1	5.784
Propane	237.1	5.118
n-Propyl alcohol	576.7	4.549
Propylene	298.9	4.678
Sulfur dioxide	335.4	4.112
Water	809.1	2.641

[a] R. A. Svehla [331].
[b] The value of σ for helium was calculated from relationships of quantum mechanics.

they may be estimated by the empirical equations

$$\frac{kT}{\varepsilon} = 1.30\frac{T}{T_c}, \tag{1.17}$$

$$\sigma = 1.18V_b^{1/3}, \tag{1.18}$$

where T_c is the critical temperature (°K) and V_b is the molar volume (cm^3/g-mol) at the normal boiling temperature. The volume V_b, in turn, may be estimated by Kopp's law of additive atomic volumes, using the values given in Table 1.4.

Table 1.4 Additive (Atomic) Volume Increments[a] for the Estimation of the Molal Volume (V_b) at the Normal Boiling Temperature

Carbon	14.8
Hydrogen	3.7
Oxygen, generally	7.4
in methyl esters and ethers	9.1
in ethyl esters and ethers	9.9
in higher esters and ethers	11.0
in acids	12.0
joined to S, P, N	8.3
Nitrogen, doubly bonded	15.6
in primary amines	10.5
in secondary amines	12.0
Bromine	27.0
Chlorine	24.6
Fluorine	8.7
Iodine	37.0
Sulfur	25.6
Ring, three membered	−6.0
four-membered	−8.5
five-membered	−11.5
six-membered	−15.0
naphthalene	−30.0
anthracene	−47.5

[a] The additive-volume method of obtaining V_b should not be used for simple molecules. The following approximate values may be employed in the estimation of σ by Equation 1.18: H_2, 14.3; O_2, 25.6; N_2, 31.2; air, 29.9; CO, 30.7; CO_2, 34.0; SO_2, 44.8, NO, 23.6; N_2O, 36.4; NH_3, 25.8; H_2O, 18.9; H_2S, 32.9; COS, 51.5; Cl_2, 48.4; Br_2, 53.2. However, the values of σ given in Table 1.3 are preferable if listed.

For diffusion in binary systems, ε_{12} and σ_{12} are estimated by the empirical expressions

$$\varepsilon_{12} = \sqrt{\varepsilon_1 \varepsilon_2}\,; \qquad \frac{kT}{\varepsilon_{12}} = \frac{kT}{\sqrt{\varepsilon_1 \varepsilon_2}}, \tag{1.19}$$

$$\sigma_{12} = \tfrac{1}{2}(\sigma_1 + \sigma_2). \tag{1.20}$$

Diffusion coefficients can be estimated by Equation 1.16 with considerable assurance. Calculated values of D_{12} have been shown to differ from 114 experimental values for a number of binary systems and a wide range of temperatures by an average of 8% [264]. This comparison was made with ε and σ estimated from T_c and V_b; better results are obtained when ε and σ are available from viscosity data. The temperature function included in Equation 1.16 is well supported by the few high-temperature diffusion data.

Equation 1.16 indicates that D_{12} is inversely proportional to the total pressure. This is approximately true up to roughly 20 atm at any temperature, and to several hundred atmospheres at high temperatures ($T/T_c > 3$). No satisfactory method of predicting D_{12} at reduced temperatures of 1 to 1.5 and reduced pressures above 0.5 is available; the best is that of Slattery and Bird [324]. Equation 1.16 does not apply to diffusion in pores at very low pressures; Knudsen diffusion is independent of pressure.

Example 1.1 Estimation of Diffusion Coefficient in a Binary Gas Mixture

Using Equation 1.16, estimate D_{12} for thiophene in hydrogen at 660°K (730°F) and 30 atm (425 psig). (Thiophene is taken as an example since it is representative of the organic sulfur compounds that are hydrogenated in the commercial hydrodesulfurization of petroleum naphtha.)

From Table 1.4, V_b for thiophene is $4 \times 14.8 + 4 \times 3.7 + 25.6 - 11.5 = 88.1$. From Equations 1.18 and 1.20 and the value 2.827 for σ for hydrogen from Table 1.3, σ_{12} is 4.04. T_c for thiophene is estimated [264] to be 579°K. Using $\varepsilon/k = 59.7$ for hydrogen (Table 1.3), kT/ε_{12} is found from Equations 1.17 and 1.19 to be 4.04. Table 1.2 then gives Ω_D as 0.8817. Substituting in Equation 1.16, we have

$$D_{12} = \frac{(0.001858)(660)^{3/2}[(2+84)/(2\times 84)]^{1/2}}{(30)(4.04)^2(0.8817)} = 0.052\ \text{cm}^2/\text{sec}.$$

The diffusion coefficient is not an exponential function of $1/T$, and Arrhenius graphs do not give straight lines. Apparent activation energies can be calculated from slopes of the curves: for $\varepsilon_{12}/\boldsymbol{k}$ of 100, the slopes correspond to activation energies of 0.75 kcal at 273°K and 3.7 kcal at 1000°K. For $\varepsilon_{12}/\boldsymbol{k}$ of 500, the corresponding values are 1.0 and 3.6 kcal. However, D_{12} is well represented by a power function of temperature, the exponent being 1.65 for $\varepsilon_{12}/\boldsymbol{k} = 100$ and 1.82 for $\varepsilon_{12}/\boldsymbol{k} = 500$ (200–5000°K).

1.4 Diffusion in Liquids

The molecular theory of liquids is relatively undeveloped and it is not possible to treat diffusion in liquid systems with the same rigor as diffusion in gases. Two complicating factors are the considerable variation of D_{12} with concentration, and the fact that mass density is more nearly constant than molar density in most instances.

In technical calculations it is customary to employ Equation 1.1 and appropriate integrated forms obtained by assuming D_{12} independent of concentration. Equation 1.12 for steady-state diffusion of a single solute is written

$$N_1 = \frac{D_{12}}{x_0}[(c_1)_1 - (c_1)_2]. \tag{1.21}$$

If the mass density is constant and Equation 1.1 is assumed to apply for diffusion relative to the plane of no net *mass* transfer, then Equation 1.10 becomes

$$\frac{\partial c_1}{\partial t} + (\boldsymbol{u}_m \cdot \boldsymbol{\nabla})c_1 = D_{12}\nabla^2 c_1 + Q_1. \tag{1.22}$$

The theory of diffusion in multicomponent liquid systems has not been developed except for dilute aqueous solutions of electrolytes [349].

Empirical correlations have been developed which may be used to estimate values of D_{12} for dilute solutions of the diffusing solute. One of the most widely used of these is that of Wilke and Chang [390]:

$$D_{12} = 7.4 \times 10^{-10}\,\frac{T(XM_2)^{1/2}}{\mu V_b^{0.6}}, \tag{1.23}$$

where D_{12} is obtained in cm^2/sec, T is the temperature in °K, X is an empirical "association parameter" of the solvent, M_2 is the molecular weight of the *solvent*, μ is the viscosity of the solution in *poises*, and V_b is the molar volume of the diffusing *solute* in cm^3/g-mol, obtained from Kopp's law (see Table 1.4). The parameter X is taken as 2.6 for water,

1.9 for methanol, 1.5 for ethanol, and 1.0 for benzene, ether, heptane, and all other unassociated solvents. The average error in the use of Equation 1.23 for the estimation of D_{12} in aqueous systems is about 10–15 per cent; considerably greater errors are common in the case of organic solvents.

Equation 1.23 is not dimensionally consistent; the variables with the specified units must be employed. Furthermore, Equation 1.23 does not hold for the diffusion of large molecules, for which the Stokes-Einstein equation gives

$$D_{12} = \frac{1.05 \times 10^{-9} T}{\mu V_b^{1/3}}. \tag{1.24}$$

Values of D_{12} calculated by the use of Equation 1.23 should not be used if they are larger than given by Equation 1.24, i.e., if $V_b > 0.27 (XM_2)^{1.87}$.

Example 1.2 Estimation of Diffusion Coefficient in a Liquid System

Estimate D_{12} for thiophene in dilute solution in hexane at 40°C.

Hexane is presumed not to associate, X is taken as 1.0, M_2 is 86 (hexane), and V_b was found in the previous numerical example to be 88.1 (thiophene). The viscosity of hexane at 40°C is 0.262 cP. Substituting in Equation 1.23, we obtain

$$D_{12} = 7.4 \times 10^{-10} \frac{313(86)^{1/2}}{0.00262(88.1)^{0.6}} = 5.6 \times 10^{-5} \text{cm}^2/\text{sec}.$$

Since V_b is evidently much less than $0.27 (XM_2)^{1.87}$, Equation 1.23 should apply.

Perhaps the most significant feature of D_{12} for liquid systems is that at ordinary temperatures it is of the order of 10^{-4} times D_{12} for typical gas systems at atmospheric pressure, or one per cent of D_{12} for gas systems at 100 atm.

Equation 1.23 suggests that D_{12} is proportional to T/μ and since liquid viscosity decreases with increase in temperature, D_{12} increases roughly as the square of the absolute temperature. As for gases, the activation energy is much smaller than for most chemical reactions. For solutes in hexane, for example, the apparent activation energy over the range 0–70°C is 2.23 kcal/g-mol. Data quoted by Maxwell [208] show T/μ for a 43.2° A.P.I. Pennsylvania crude fraction boiling from 250–275°C (482–527°F) to increase as the 4.4 power of T. At 500°K (440°F) this corresponds to an activation energy of 4.3 kcal/g-mol.

Table 1.5 is an abbreviated list of experimental values of D_{12}° for low solute concentrations in common liquid systems. Himmelblau [144] presents a detailed review on diffusion of dissolved gases in liquids. Diffusion in general is treated in the books by Sherwood and Pigford [319], by Bird, Stewart, and Lightfoot [32], and by Jost [163] and in a review by Bird [31]. Methods of estimating diffusion coefficients are summarized and evaluated in the book by Reid and Sherwood [264]. Crank [79] gives the mathematical solutions to the diffusion equations for a variety of geometrical shapes and boundary conditions.

Table 1.5 Diffusion Coefficients for Binary Liquid Systems at Low Solute Concentrations
(Experimental values of D_{12}° (cm^2/sec) $\times 10^5$)

	Solute	T(°K)	$D_{12}^{\circ} \times 10^5$
In Water	Helium	298	6.3
	Hydrogen	298	4.8
	Oxygen	298	2.41
	Carbon dioxide	298	2.00
	Ammonia	285	1.64
	Chlorine	298	1.25[a]
	Methane	293	1.49
	Propane	277	0.55
		333	2.71
	Propylene	298	1.44
	Benzene	298	1.09
	Methanol	288	1.26
	Ethanol	283	0.84
	n-Propanol	288	0.87
	n-Butanol	288	0.77
	i-Butanol	288	0.77
	i-Pentanol	288	0.69
	Ethylene glycol	293	1.04
	Glycerol	293	0.82
	Acetic acid	293	1.19
	Benzoic acid	298	1.21
	Glycine	298	1.05
	Ethyl acetate	293	1.00
	Acetone	288	1.22
	Furfural	293	1.04
	Urea	293	1.20
	Diethylamine	293	0.97
	Aniline	293	0.92
	Acetonitrile	288	1.26
	Pyridine	288	0.58

[a] Equilibrium mixture of hydrolyzed and unhydrolyzed chlorine.

Table 1.5—*Continued*

	Solute	T(°K)	$D_{12}^{\circ} \times 10^5$
In Benzene	Acetic acid	298	2.09
	Carbon tetrachloride	298	1.92
	Ethylene chloride	281	1.77
	Ethanol	288	2.25
	Methanol	298	3.82
	Naphthalene	281	1.19
In Acetone	Acetic acid	298	3.31
	Benzoic acid	298	2.62
In Ethanol	Carbon dioxide	290	3.20
	Pyridine	293	1.10
	Urea	285	0.54
	Water	298	1.13
In Toluene	Acetic acid	293	2.00
	Acetone	293	2.93
	Benzoic acid	293	1.74
	Ethanol	288	3.00

1.5 Solid Catalysts

Industrial catalysts comprise a wide variety of materials and are manufactured by a variety of methods. Many catalysts or porous structures studied in the laboratory in fundamental investigations are chosen so as to have simple, uniform, or known structure rather than high activity or good mechanical strength. Therefore, they are frequently of little immediate industrial interest.

The commercially useful catalyst particle size is determined by the process in which it is to be used. For fixed beds particles generally range from about $\frac{1}{16}$–$\frac{1}{2}$ in. in diameter. Diffusional resistance within the porous structure increases with particle size and the large internal pore surface becomes less "effective," so particles larger than about $\frac{1}{2}$ in. are frequently pierced with holes or formed as rings. Sizes much smaller than $\frac{1}{16}$ in. may produce excessive pressure drop through the bed, be mechanically weak, or difficult to manufacture. In fluidized-bed reactors a substantial particle-size distribution is desired for good fluidization characteristics; this is usually present in the catalyst powder supplied, or is produced in the normal course of operation by the gradual attrition of the catalyst in the highly turbulent environment. The particles present generally range from about 20–300 μ in diameter, the mean being about 50–75 μ. Smaller particles are entrained from the reactor; larger particles fluidize poorly. Catalysts for slurry reactors are typically 75–200 μ in diameter. Finer powders are difficult to remove by settling or filtration; coarser powders may be more difficult to suspend and may be less effective per unit mass.

The simplest method of catalyst manufacture and one widely used is to impregnate a catalyst support with an appropriate solution, followed by drying and various treatments such as reduction or calcination, to produce an active catalyst. The pore structure of the final catalyst is essentially that of the support (carrier) but the concentration of the active catalyst usually decreases toward the center of the particle. The degree of uniformity varies with the adsorptive properties of the carrier and the method of manufacture. Thus, the use of an alcoholic solution may produce a substantially different concentration distribution than that obtained with an aqueous solution. The drying following impregnation by a solution usually takes place by evaporation at the surface of the pellet, to which the solution is drawn by capillarity. The solute tends to be precipitated near the surface as the solvent evaporates. If the active ingredient is expensive, as with noble metals such as platinum or palladium, it is sometimes desired to support it in the form of a thin annular shell on the outside of the pellet if the reaction is so rapid that active material in the particle interior would contribute little. With complex reactions, selectivity is usually diminished in the presence of significant concentration gradients through a porous catalyst (low effectiveness factor). Confining the active catalyst to a thin outside layer provides a method of eliminating this problem while retaining a catalyst particle of a size easy to work with.

Alternatively, many industrial catalysts are prepared by a process starting with precipitation from aqueous solution, sometimes in the presence of finely powdered carriers, followed by dewatering and drying. The resulting solid powder is then typically mixed with binders and lubricants, the mixture is pelletized or extruded, and the pellets or extrudates are subjected to a heat treatment or "activation" at high temperatures, which has several purposes. It is usually desirable to decompose the inorganic material present, such as a nitrate or a carbonate, to form the corresponding oxide. In most cases these reactions also cause fine pores to be opened up in the structure with a corresponding increase in total surface area. It is also desirable to eliminate components undesired in the final composition, such as organic binders and die lubricants, by burning them out with air. Sometimes a partial sintering is required to increase mechanical strength, although too much sintering as caused by excessively high temperatures may decrease the effective diffusivity in the structure (see, for example, Table 1.11). It is usually desirable to heat the catalyst during manufacture to a temperature at least equal to the highest reaction temperature to which it will be ultimately subjected, so that it will be structurally stable in the reactor. This method of manufacture usually results in a uniform distribution of the active catalyst throughout the pellet.

Silica-alumina cracking catalysts are prepared from a hydrogel, frequently incorporating fine crystals (about a micron in size) of a zeolite (molecular sieve). For use in moving-bed processes, the hydrogel is formed into "beads" by dispersal as drops of controlled size into hot oil which coagulates the gel. For use in fluidized-bed reactors a "microspheroidal" catalyst is usually used, made by drying the hydrogel in a spray tower.

1.5.1 *Catalyst Supports*

The most ubiquitous catalyst carrier is alumina. It is inert to most reacting systems, structurally stable to relatively high temperatures and is available in a variety of forms with surface areas ranging from less than 1 m^2/g up to about 300 m^2/g. Silica gel, either in the form of granules or powder, is also a useful carrier and is available with surface areas up to about 800 m^2/g. If the catalyst is to be composited, one may start with powdered silica gel, with a hydrogel, with colloidal silica, or utilize a hydrogel formed *in situ* by precipitation. Diatomaceous earth (kieselguhr), a naturally occurring form of silica having surface areas in the neighborhood of 50 m^2/g, may be utilized as a powdered carrier. Porous carbon is thermally stable to temperatures of 1000°C or more under inert conditions and certain forms have the highest known surface areas of any material, up to about 1300 m^2/g. These "activated" carbons are commonly used as catalyst carriers for organic reactions, as in slurry reactors.

For high-temperature reactions it may be necessary to have a "refractory support" for mechanical stability. These are primarily various forms of silica and alumina manufactured by high-temperature fusion (e.g., 2000°C) in an electric furnace. The product is then crushed and sieved, formed into irregular granules, spheres, rings, cylinders, etc., and fired in a kiln typically at about 1400°C. The final products have surface areas of less than 1 m^2/g and pore sizes typically in the range of 20–100 μ. These supports are also useful where high reaction rate per unit mass of catalyst is less important than other factors such as catalyst cost, or where the desired product is a reaction intermediate and it is necessary to eliminate fine pores in order to enhance reaction selectivity.

1.5.2 *Zeolites* ("*Molecular Sieves*")

A major change in cracking and hydrocracking processes in petroleum refining has been brought about in the last few years by the introduction of zeolites as catalysts. Zeolites are highly crystalline hydrated alumino-silicates which upon dehydration develop in the ideal crystal a uniform pore structure having minimum channel diameters of from 3–10 Å depending upon the type of zeolite and the nature of the cations present. The structure consists of a three-dimensional framework of SiO_4 and AlO_4 tetrahedra, each of which contains a silicon or aluminum atom in the center. The oxygen atoms are shared between adjoining tetrahedra. The two types of tetrahedra can be present in various ratios and arranged in a variety of ways. Zeolites have the general formula

$$Me_{2/n}O : Al_2O_3 : X\,SiO_2 : Y\,H_2O,$$

where Me represents a metal cation of valence n and X and Y, which exceed unity, vary with the type of zeolite. The metal cation is present because for each alumina tetrahedron in the lattice there is an over-all charge of -1. This requires a cation to produce electrical neutrality.

Since the pore sizes of zeolites are closely comparable to the size of the molecules of many substances of industrial interest, zeolites can exhibit a selective form of adsorption based upon the exclusion from the pores of molecules whose size or shape cannot permit them to be accommodated, hence the term "molecular sieves." They have been used for over a decade in separation processes. To achieve high catalytic activity in acid-catalyzed reactions such as cracking, the sodium cations originally present in the zeolite are replaced by other cations such as those of rare earths and by hydrogen to produce acid sites. The zeolites of interest in catalysis are primarily those having relatively large pores; the largest pores are exhibited by faujasite-type zeolites, which are characterized by a three-dimensional structure of cages or cavities inter-connected by somewhat smaller portholes. Types 13X and 10X are, respectively, the sodium and calcium forms of a synthetic zeolite having a structure identical to the natural mineral faujasite and a Si/Al atomic ratio of about $1\frac{1}{4}$. Type Y has a structure similar to natural faujasite, but an Si/Al atomic ratio of about $2\frac{1}{2}$. Both X and Y can be prepared within certain ranges of Si/Al ratios. In the sodium form of Type X or Y the cages are about 16 Å in diameter and the interconnecting port-holes about 10 Å diameter. For 10X the portholes are about 8 Å in diameter. The next largest pores are found in mordenite, which has a two-dimensional pore structure consisting of parallel slightly elliptical straight channels, having about 7 and 6 Å major and minor diameters, respectively, in the sodium form, the channels being effectively isolated one from another. The smaller pore synthetic zeolites include 4A and 5A, which are the sodium and calcium forms of Type A, with pore diameters of about 4 and 5 Å, respectively. The structures of zeolites in general are described in a book by Hirsch [145], a review by Breck [45], and in a recent symposium [45].

Synthetic faujasite-type zeolites are generally manufactured in the form of crystals about 1 μ in size. For use in cracking or hydrocracking processes 5–15 per cent of the zeolite is usually incorporated in a silica-alumina matrix, like that comprising conventional silica-alumina catalysts, with powdered α-alumina sometimes added for attrition resistance, and the mixture is then formed into beads, powders, etc., of the desired size.

1.5.3 *Physical Characterization of Catalysts*

The ultimate goal here is to be able to predict the effective diffusivity of reactants and products in a porous catalyst. The closely related problem, that of predicting the permeability of porous media to flow of fluids, has been extensively studied for many years [66]. Porous structures are too complicated and too various to be capable of representation by a single number or even one simple model. Prediction must be based on enlightened empiricism, combining theoretical models and physical measurements on catalysts with empirical information on method of manufacture and the probable effects of phenomena occurring during use in reactors.

Knowledge of the total surface area of the catalyst is a basic requirement. This is generally obtained by the B.E.T. (Brunauer-Emmett-Teller) method, in which the effect of the total pressure on the amount of a gas adsorbed on the solid at constant temperature is measured. The experimental procedure and the methods of analysis have been developed in great detail. For the most reliable measurements the molecules of the gas chosen should be small, approximately spherical, inert so no chemisorption takes place, and the gas should be easy to handle at the required temperature. The choice is usually nitrogen; measurements are required over a range of relative pressure P/P_0 (ratio of gaseous pressure to vapor pressure over liquid phase at the same temperature) of about 0.05 to 0.3 and, therefore, are made at cryogenic temperatures, usually using liquid nitrogen as the coolant.

The various methods of analyzing the data are all directed to determination of the quantity of adsorbate that corresponds to a molecular monolayer on the solid surface. Combining this with the cross-sectional area occupied per adsorbed molecule, usually taken to be about 16.2 Å^2 for nitrogen, gives the area of the catalyst. When the total area of a sample is small, the amount of nitrogen gas adsorbed becomes small relative to the total amount in the apparatus and the accuracy of measurement becomes poor. By using as an adsorbate a vapor of higher boiling point, measurements at liquid nitrogen temperature can be made at much lower pressures to achieve the desired range of P/P_0 values, so the amount of gas adsorbed on the solid is now a much larger fraction of that present and can be more precisely measured. The gas most frequently used for such lower pressure measurements is krypton, which has a vapor pressure of about 3 mm Hg at the temperature of liquid nitrogen.

Pore-Size Distribution. The most important requirement for predicting diffusivity is a knowledge of the pore-size distribution of the catalyst. In its absence, it is impossible to make more than a gross guess as to the effective diffusivity unless the diffusion is completely by the bulk mode, a condition encountered infrequently in gas-phase reactions. Much literature on pore diffusion effects in catalysts, particularly the earlier literature, provides insufficient information on the pore structure to permit interpretation in more than a qualitative manner.

The starting point is to assume that the complicated pore geometry can be represented by a simple model, usually an array of cylindrical capillaries of uniform but different radii, randomly oriented. The fact that pore cross sections are actually highly irregular is not of major consequence in the application of theory. The distribution of fine pores is usually determined by a method developed by Barrett, Joyner, and Halenda [21] and improved by Cranston and Inkley [80]. It applies the Kelvin equation, which relates vapor pressure above a liquid in a capillary tube to its surface curvature. The increase in amount of vapor adsorbed onto a catalyst upon an incremental increase in vapor pressure at constant temperature (usually nitrogen vapor at the temperature of liquid nitrogen) represents the filling of capillaries of a size given by the Kelvin equation. This must be corrected for an increase in thickness of the adsorbed layer. The size of the largest pores that can be measured is limited by the rapid change of meniscus radius with pressure as the relative pressure, P/P_0, nears unity. This is generally taken as about 300 Å in diameter, corresponding to a relative pressure of 0.93. The smallest pore sizes that can be determined by this method are about 15–20 Å in diameter. Although measurements may be reported corresponding to smaller pore sizes, interpretation of the results becomes increasingly uncertain. The method of analysis assumes that the properties of the condensed phase in the capillaries are the same as those of a bulk liquid, yet the concepts of surface tension or a curved surface must become increasingly unrealistic as pore size becomes of the order of magnitude of the size of adsorbate molecules. A somewhat different relationship between amount of vapor sorbed and pressure is usually obtained experimentally upon decreasing rather than increasing the pressure, and an extensive literature exists attempting to relate these hysteresis effects to details of the pore structure or to explain them in terms of supersaturation effects, variation of contact angle, etc. These, however, are secondary matters which do not greatly affect the use of pore-size-distribution measurements for our purpose. The desorption curve is preferred and good agreement has generally (although not always) been found between the pore-size-distribution curve by nitrogen desorption and that determined by mercury porosimetry, described below. Studies by Joyner, Barrett, and Skold [164] showed this to be true for a variety of charcoals; indeed, a double peak in one sample having maxima at about 25 and 75 Å was traced out by both methods. The surface area of the catalyst can also be determined from integration of all the pore volumes filled up as vapor is adsorbed and condensed, which provides an independent measurement of the surface area from that determined by the usual B.E.T. method.

A simple application of these same principles was developed by Benesi, Bonnar, and Lee [27]. They use a mixture of a nonvolatile and a volatile liquid (e.g., carbon tetrachloride and cetane (n-$C_{16}H_{34}$)) at room temperature to obtain the desired vapor pressure. A dried catalyst sample can be placed in a desiccator having the solution in the bottom and the gain in weight combined with the known density of carbon tetrachloride gives the cumulative pore volume up to a critical diameter corresponding to the relative pressure in the desiccator. By varying the ratio of carbon tetrachloride to cetane in the liquid the partial pressure of carbon tetrachloride can be varied, thus providing information on pore-size distribution.

Mercury Porosimetry. Mercury does not wet most surfaces, so an external pressure is required to force it into a capillary. The relationship is

$$r = (-2\sigma \cos \theta)/P, \tag{1.25}$$

where σ is the surface tension of mercury, and θ is the contact angle with a surface. Ritter and Drake [270] found that the contact angle between mercury and a wide variety of materials such as charcoals and metal oxides varied only between 135° and 142° and suggested that an average value of 140° could be used in general. Equation 1.25 thus reduces to

$$r = 75\,000/P, \tag{1.26}$$

where r is pore radius (Å) and P is in atmospheres. The increase in amount of mercury forced into the porous material upon an increase in pressure represents the filling of pores of a size given by Equation 1.26. The smallest pore sizes that can be detected depend upon the pressure to which mercury can be subjected in a particular apparatus. Pore diameters down to about 200 Å can be determined with available commercial apparatus, and high-pressure porosimeters have been built to measure pores down to 30 Å diameter. Pores larger than about 7.5 μ (75 000 Å) will be filled at atmospheric pressure. Other investigators and laboratories have used somewhat different values of contact angle and surface tension, resulting in values of $(r \cdot P)$ varying from 75 000 down to as low as 60 000, which can result in as much as a 25 per cent variation in reported pore size. The most commonly used value at present is about 62 000.

Void Fraction. The total pore volume of a catalyst can be determined simply by measuring the increase in weight when the pores are filled with a liquid of known density. The liquid should preferably be of low molecular weight so that fine pores are filled; water, hydrocarbons, or chlorinated hydrocarbons may be used satisfactorily. A simple procedure is to boil a sample of dry pellets of known weight in distilled water for about 30 min, replace the hot water with cool water, transfer to a damp cloth, roll to remove excess water and reweigh. This will determine total pore volume between about 10 and 1500 Å. The accuracy, however, is limited by the difficulty of drying the external surface without removing liquid from large pores, and some liquid tends to be held around the points of contact between particles.

More accurate results are obtained by the so-called mercury-helium method. A container of known volume, V cm³, is filled with a known weight of pellets or powder, W g. After evacuation helium is admitted, and from the gas laws is calculated the sum of the volume of the space between the pellets V' plus the void volume inside the pellets V_g. The true density of the solid is then

$$\rho_t = \frac{W}{V - (V' + V_g)}. \tag{1.27}$$

The helium is then pumped out and the bulb filled with mercury at atmospheric pressure. Since the mercury does not penetrate the pores, its volume is that of the space between the pellets V'.

The porosity or void fraction θ (cm^3/cm^3) is given by

$$1 - \theta = \frac{V - (V' + V_g)}{V - V'}. \tag{1.28}$$

The density of the pellets $\rho_p = W/(V - V')$ (g/cm^3).

The void fraction determined from the helium volume is sometimes slightly greater than that determined by absorption of a liquid, since the smaller volume of the helium molecule permits it to penetrate into fine pores inaccessible to larger molecules.

More details concerning methods for physical characterization of porous substances are given in the book by Gregg and Sing [124]. The proceedings of a symposium on structure and properties of porous materials also contain much useful information [102].

1.5.4 *Bimodal Pore-Size Distribution*

Many catalysts or porous substances show a bimodal pore-size distribution, sometimes termed a *bidisperse*, or *macro-micro*, distribution. This is the case, for example, of catalyst pellets prepared in the laboratory by compacting fine porous powders. One then deals with a fine pore structure within each of the particles of the original powder, plus a coarser pore structure formed by the passageways around the compacted particles. As compacting pressure is increased the micropores remain unaffected unless or until the crushing strength of the solid is exceeded, but the macropores become successively reduced in size. As an example, boehmite pellets pressed in the laboratory by Otani, Wakao, and Smith [242] showed a micropore size distribution with a maximum at about 20 Å radius, unaffected by degree of applied pressure, and macropore size distribution maxima varying from 5000–500 Å radius with increased pressure. A bimodal pore-size distribution is also found in most commercial forms of alumina. That used by Rothfeld [278] showed two pore-size-distribution peaks, one at 1.25 μ (12 500 Å) and one at 120 Å (diameters). About 65 per cent of the total pore volume and 99 per cent of the surface was in the micropores. A nickel oxide-on-alumina catalyst prepared by Rao and Smith [259] showed pore-size-distribution peaks at about 30 and 1000 Å radius. The fine particles themselves may also show a bimodal-type distribution. Thus, an examination by electron microscopy of a sample of commercial alumina by Bowen, Bowrey, and Malin [42] seemed to show the presence of three kinds of pores: (1) large irregular spaces *between* discrete particles, these particles being of the order of 2000 Å diameter, (2) smaller irregular holes *inside* the particles, probably representing residual space between agglomerates which came together to form the particle, and (3) well-ordered micropores approximately cylindrical in cross section having a mean diameter of 27 ± 1 Å. In this case, about 18 m^2/g of the total area of about 275 m^2/g was associated with the large macropores. Many commercial porous carbons also have a bimodal pore-size distribution, as do catalysts prepared by incorporating molecular sieves in a gel matrix or pressing them into pellets. The pore-size distribution into clearly defined macropores and micropores is usually more marked in laboratory pressed powders than in most commercial catalysts. Where a clear-cut division does not exist, the micropore region is frequently defined, somewhat arbitrarily, as comprising pores under about 200–250 Å diameter. Some industrial catalysts are deliberately made to have a macro-micropore system in order to minimize or eliminate diffusion limitations. An example is a recently patented catalyst for ammonia synthesis in which the catalytic material, a promoted iron oxide, is crushed into a fine powder and then formed into pellets. The catalyst must be reduced before use and the existence of a network of macropores allows the water vapor formed by reduction to diffuse out of catalyst pellets more rapidly. Studies by Nielsen and co-workers [Reference 227, pp. 97–100 and p. 162] show that loss of surface area is thereby minimized and a catalyst of higher intrinsic activity is obtained. An ammonia synthesis catalyst structure of this type is of more significance if the catalyst is to be reduced at relatively low pressures where diffusion is in the Knudsen or transition range where effective diffusivity is a function of pore size (Section 1.7).

1.5.5 *Measurement of Diffusion in Porous Solids*

The most common method of measuring counterdiffusion rates is by a steady-state technique apparently first used by Wicke and Kallenbach [387] and modified by Weisz [372, 377, 378], and by Smith and co-workers [140]. Two pure gases are allowed to flow past opposite faces of one or more cylindrical pellets affixed in parallel in a tight-fitting mount such as plastic tubing or a gum rubber disk. The fluxes through the pellet are calculated from a knowledge of the gas flow rates and composition leaving each side of the diffusion cell containing the pellets. No significant pressure gradient is allowed to develop across the pellet. Typical gas pairs are hydrogen-nitrogen, or helium-nitrogen. Since this method measures diffusion *through* the pellet it ignores dead-end pores, which may contribute to reaction and are included in a measured pore-size distribution. It may be used over a wide range of pressures but cannot be readily adapted to temperatures much exceeding ambient. It is primarily applicable to well-formed cylindrical pellets, although Weisz has used it for spheres by applying a geometrical correction factor.

Steady-state flow under a pressure gradient gives results difficult to interpret unless all of the pores are sufficiently small that only Knudsen flow occurs. In the time-lag method of Barrer, one side of the pellet is first evacuated and then the increase in this downstream pressure is observed with time, the upstream pressure being held constant. The change in pressure drop across the pellet during the experiment is held to an insignificantly small value. There is a time lag before a steady-state flux develops, and effective diffusion coefficients can be calculated from either the unsteady-state or the steady-state data. Under unsteady-state conditions correct analysis must allow for accumulation or depletion of material by adsorption, if this occurs, even if surface diffusion is insignificant (see Equations 1.46 and 1.47).

Several unsteady-state techniques have been developed for the measurement of pore diffusion. In one group of methods the solid is first saturated with one fluid and then the loss in weight is followed as a function of time after a vacuum is suddenly applied, or a second carrier gas is suddenly allowed to flow over the solid and the concentration of the leaving gas is followed with time [122]. Most of the interchange in a spherical pellet occurs in a time $t = r^2/10D_{eff}$ where r is the pellet radius. For typical catalyst pellets and gases this time is so short as to make it difficult to obtain accurate results (e.g., for $r = 0.3$ cm and $D_{eff} = 10^{-3}$ cm^2/sec, $t = 9$ sec) and the method is primarily applicable to liquids, or very finely porous solids having low diffusivities such as crystals of molecular sieves. Deisler and Wilhelm [94] devised a sophisticated frequency response technique which, however, requires a complicated test method and involved procedures to interpret the data. A promising unsteady-state method is based on gas chromatography, interpreting the broadening of an input pulse. Only preliminary and scattered results have been reported and a detailed assessment of its degree of validity has not yet been made, but it provides a potential method of studying a variety of shapes and of making measurements under experimental conditions such as high temperatures that are not amenable to the Wicke-Kallenbach method. Surface diffusion, adsorption phenomena, and unsteady-state phenomena may make it difficult to interpret results (see Section 1.7.3), and the proper method of analyzing data from pellets having a wide pore-size distribution is not yet clear. Nevertheless, if an accurate effective diffusion coefficient can be obtained by this method, which relates to fluxes in and out of the particle, it may be more representative of the situation occurring in reaction than steady-state diffusion measurements *through* a porous body, whenever the catalysts are highly anisotropic. The effective diffusivity may also be determined from measurements under diffusion-limited conditions of the rate of burnoff of carbon from a uniformly coked catalyst (see Section 5.3.2).

Dead-end pores, if they are present, will contribute to the flux in unsteady-state methods, but not in steady-state measurements. Davis and Scott [91] used both methods to study diffusivities in spheres of activated alumina and a Norton catalyst support. After eliminating a skin containing fine pores, good agreement was found between results for the two methods, which indicates that dead-end pores were unimportant for these catalyst supports.

An effective diffusion coefficient may also be determined from reaction rate measurements on catalyst pellets of two or more sizes by the methods of Chapters 3 and 4, provided that the mathematical form of the intrinsic kinetics of the reaction is well established. For simple power-law expressions it is not necessary to know the rate constant.

In any of these methods it is important that the data be analyzed by the proper flux equation. This point may be illustrated with reference to the Wicke-Kallenbach method. The process of counterdiffusion changes as the diameter of the pores becomes progressively smaller. When the channels are large (e.g. above about 10 μ at 1 atm) equimolar counter-diffusion occurs and $D_{12,\mathrm{eff}}$ can be calculated directly from Equation 1.14 if the flux is known. As the pore size is progressively reduced, the relative rates of the counterdiffusing gases gradually change, ultimately becoming inversely proportional to the square root of the molecular weight ratio of the two gases. The proportionality of N_1/N_2 to $\sqrt{M_2/M_1}$ occurs well before the point at which the pore radius approaches the mean free path. This has been demonstrated both theoretically and experimentally by Scott and co-workers [311, 312]. In principle, the relationship between flux ratio and molecular weight ratio can be employed to calculate effective diffusivities from measurements on only one of the two gas fluxes. In practice, however, measurements of both fluxes are highly desirable. Deviations from the theoretical ratio may indicate experimental difficulties such as leaks or the existence of surface diffusion.

1.6 Bulk Diffusion in Porous Catalysts

Pore diffusion may occur by one or more of three mechanisms: ordinary diffusion, Knudsen diffusion, and surface diffusion. If the pores are large and the gas relatively dense (or if the pores are filled with liquid), the process is that of bulk, or ordinary, diffusion, which has been discussed in preceding sections. If the pores are randomly oriented the mean free cross section of the porous mass is the same in any plane and is identical with the volume fraction voids θ. If the pores were an array of cylinders parallel to the diffusion path, the diffusion flux per unit total cross section of the porous solid would be the fraction θ of the flux under similar conditions with no solid present. However, the length of the tortuous diffusion path in real pores is greater than the distance along a straight line in the mean direction of diffusion. Moreover, the channels through which diffusion occurs are of irregular shape and of varying cross section; constrictions offer resistances that are not offset by the enlargements. Both of these factors cause the flux to be less than would be possible in a uniform pore of the same length and mean radius. We may thus express a bulk diffusion coefficient per unit cross section of porous mass, $D_{12,\mathrm{eff}}$, as

$$D_{12,\mathrm{eff}} = \frac{D_{12}\theta}{L'S'} = \frac{D_{12}\theta}{\tau}, \tag{1.29}$$

where L' is a length or angle factor and S' a shape factor, both being greater than unity, to allow for the two effects. There have been many attempts to develop theoretical expressions relating L' or S' to some easily measurable property such as porosity or particle size in a compact (see summary by Masamune and Smith [200]) but none is of general applicability, although L' is always predicted to increase with decreasing void fraction.

Theoretical models of pores with constrictions to estimate the value of S' have been proposed by Petersen [248], Currie [83], and Michaels [215]. Petersen represented the pores as a series of hyperbolas of revolution with constrictions at the vertices of the hyperbolas. Currie represented the pores as tubes of sinusoidal form. Michaels represented the pore as a repetition of two cylindrical capillaries of different diameters and lengths joined in series. In each case the rate of diffusion is compared to that for a pore of uniform diameter and the same surface-to-volume ratio as that of the model. Figure 1.3 compares the three models by graphing $1/S'$ against the ratio of maximum to minimum cross section.

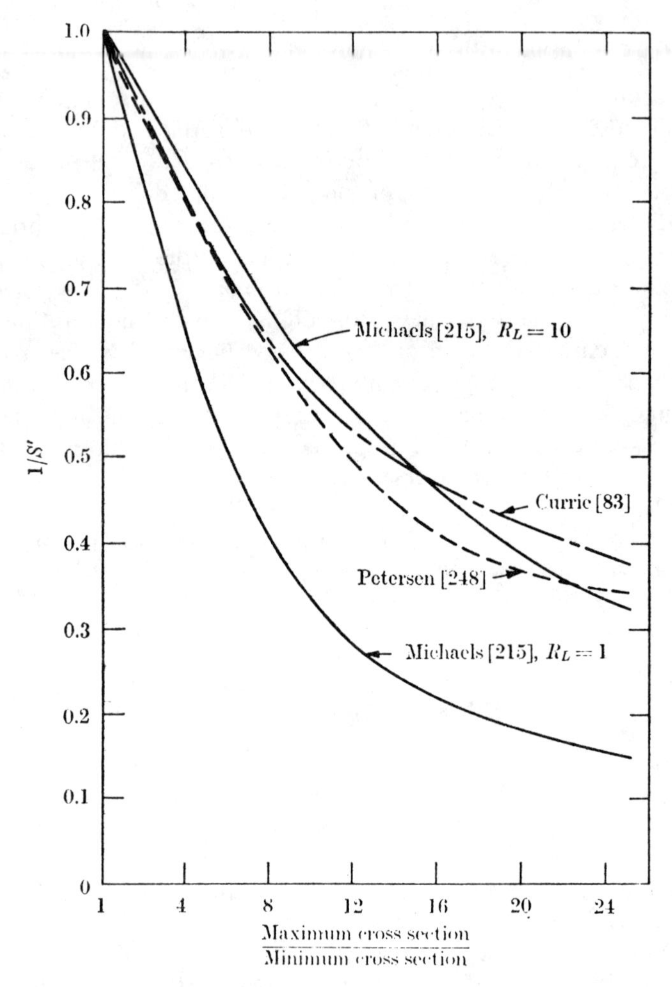

Figure 1.3. Theoretical effect of variation in pore cross section on effective diffusivity.

In Michaels' model the ratio of lengths of the two capillary cylinders in series $R_{L'}$ is a further adjustable parameter (or alternately, the ratio of their widths) and Figure 1.3 shows two extreme values. The value of S' is seen to depend primarily on the ratio of maximum to minimum pore area and is not greatly affected by the details of the model. S' can reach values of 3 to 4 with moderate degrees of necking.

For most real porous systems it is impossible to separate L' and S', so they are multiplied together here and represented by τ, a single factor to allow for both varying direction of diffusion and varying pore cross section. For accurate work τ must be determined experimentally. It will be termed here the "tortuosity factor." In some literature L' itself is termed a tortuosity factor. Petersen [251] terms $1/S'$ a constriction factor and his group (τ/σ) is the same as τ as used here. Other literature refers to a "labyrinth factor," which is the reciprocal of τ in present nomenclature. Wheeler [382, 383] proposed a model in which the pores were visualized as cylinders of one fixed diameter which intersect any plane at an average angle of 45°. Hence $L' = \sqrt{2}$. $D_{12,\text{eff}}$ relates to the flux per unit of face area of pores at any intersecting plane, which is less than the flux per unit cross section normal to the direction of diffusion in the cylindrical pore, so another factor of $1/\sqrt{2}$ is introduced. Consequently, in Wheeler's model, $\tau = 2$. Weisz and Schwartz [378] proposed a model in which $\tau = \sqrt{3}$. For diffusion through a randomly oriented system of long cylindrical pores the tortuosity factor is 3 [161, 388].

Figure 1.4 shows the data of Currie [83] for diffusion in unconsolidated beds of various powders and granular materials, illustrating the variation of $D_{12,\text{eff}}$ with porosity. For any given point on the figure, τ is the ratio of abscissa to ordinate and it is seen to vary up to a value of 2. Hoogschagen [155] reported values of τ of 1.4 to 1.6 for loose powders and beds of glass spheres having bed porosities of 0.35 to 0.43.

Catalysts are usually consolidated porous media and somewhat larger values of τ would seem to be required than those for packed beds. Figure 1.5 and Table 1.6 gather together data reported on bulk diffusion in catalyst pellets under a variety of conditions. Amberg and Echigoya [5] studied catalyst pellets prepared by pressing to various densities powders of various sizes of a silver alloy containing 8.5 per cent calcium and the same material after "activation" in which most of the calcium was removed by steam and acid treatment. (A catalyst of this composition is reportedly used for the commercial oxidation of ethylene to ethylene oxide.) The B.E.T. (nitrogen) surface areas varied from 0.22 to 1.50 m^2/g, and on the basis of forced-flow experiments diffusion was believed to be all in the bulk range. For the catalyst prepared from activated powder typical values of τ were 7.5 at θ of about 0.6, 10 at θ of 0.3, and ∞ at θ of 0.1. With powder comprising the silver alloy before activation, stated to be more brittle, τ was smaller for a given void fraction (e.g., $\tau = 6$ at θ of 0.3) and diffusion was not completely stopped ($\tau \to \infty$) until the void fraction was reduced to 0.04. The void

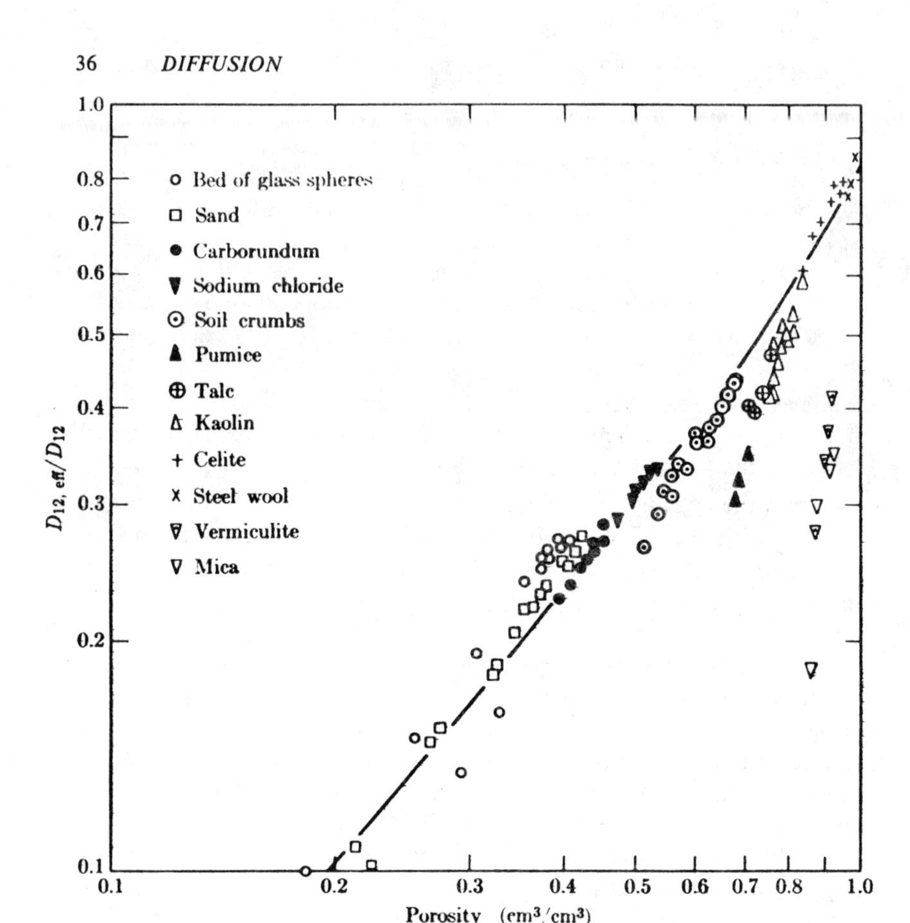

Figure 1.4. Hydrogen-air diffusion in various unconsolidated porous media. Currie [83].

fraction here includes completely enclosed pores and the greater deformability of silver than of its alloy may have caused substantial necking down of passageways and consequent increase in τ. A skin effect (compaction of the surface or filling of surface openings with very fine powder) may also have been significant, although the authors seemed to believe otherwise. Pressing of a pure commercial nickel powder gave a value of $\tau = 6$ at θ of about 0.26.

Osberg, Tweddle, and Brennan [236] prepared similar silver–8.5 per cent calcium alloy catalysts by flame-spraying onto a support. Diffusion measurements were made on these deposits after activation by autoclaving with water at 200°C, washing with acetic acid, and lifting the deposited catalysts from the support. (Resulting thicknesses were 0.09

to 0.18 cm.) These gave values of τ ranging from 2 to 3.5. Measurements also were made on disks cut from an 8.5 per cent calcium-silver ingot and leached by an electrolytic method. For θ of 0.41, τ was about 16. The residual calcium was in the range of 0.2 to 0.4 per cent. One can hypothesize that the calcium in the interior of the disk was not as completely removed as that in the rest of the disk; a thin layer of less porous

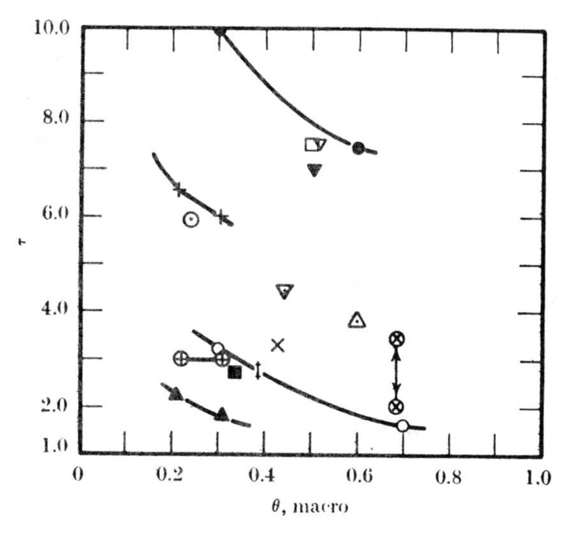

Figure 1.5. Some tortuosity factors for bulk diffusion in catalysts.

material in the center could easily give rise to a substantial resistance. Masamune and Smith [200] studied counterdiffusion of gases at atmospheric pressure through catalysts compacted from silver powder, prepared by decomposition in nitrogen of silver phthalate or other organic compounds containing silver. Evidently, a substantially different type of porous silver structure was obtained.

Saraf [290, 300] studied hydrogen-nitrogen diffusion at atmospheric pressure in tablets formed by compressing a powdered chromia-on-γ-alumina catalyst to various densities. The peak in the macropore distribution curve varied from 5000 to 30 000 Å. Diffusion was largely in the bulk range and τ varied from 1.8 at $\theta_{\text{macro}} = 0.31$, to 2.5 at $\theta_{\text{macro}} =$ 0.22, the lower void fraction corresponding to the macropore peak of 5000 Å. The studies of Wakao and Smith [355] of diffusion through boehmite (a hydrated alumina), of Saraf, and of Masamune and Smith were all of pellets formed by pressing powder into a die. The pellets had a bimodal pore-size distribution, but in all three cases it appears

Table 1.6 Tortuosity Factors for Bulk Diffusion in Catalysts (Experimental Data shown in Figure 1.5)

Catalyst	Technique	θ	τ	Reference
+ Ag–8.5% Ca alloy pelletized from powder	gas diffusion, 1 atm	0.3 0.04	6.0 ∞	Amberg and Echigoya [5]
● Ag pelletized from above powder after removal of Ca by leaching	gas diffusion 1 atm	0.6 0.3 0.1	7.5 10 ∞	Amberg and Echigoya [5]
○ Ni pelletized from commercial powder	gas diffusion, 1 atm	0.26	6	Amberg and Echigoya [5]
O Ag pelletized from powder	gas diffusion, 1 atm	0.7 0.3	1.7 3.3	Masamune and Smith [200]
⊗ sprayed Ag-alloy catalyst, after activation	gas diffusion, 1 atm	0.68 0.41	2–3.5 16 (see text)	Osberg, Tweddle, and Brennan [236]
■ pelletized boehmite alumina	gas diffusion, 1 atm	0.34 0.31	2.7 1.8	Wakao and Smith [355]
▲ pelletized Cr_2O_3/Al_2O_3 catalyst	gas diffusion, 1 atm	0.22	2.5	Satterfield and Saraf [290, 300]

Catalyst	Technique	θ	τ	Reference
□ 1% Pd on alumina spheres, commercial support	from reaction (liquid phase)	0.5	7.5	Satterfield, Pelossof, and Sherwood [298]
△ 0.5% Pd on alumina, commercial type catalyst pellets	from reaction (liquid phase)	0.59	3.9	Satterfield, Ma, and Sherwood [297]
⊕ Pelletized from 1–8 μ iron powder	gas diffusion, 1 atm	0.22–0.32	2.6–2.9	Hoogschagen [155]
▼ Harshaw commercial MeOH synthesis catalyst, prereduced	gas diffusion, 65 atm	0.49	6.9	Satterfield and Cadle [292]
× Haldor-Topsøe commercial MeOH synthesis catalyst, prereduced	gas diffusion, 65 atm	0.43	3.3	Satterfield and Cadle [292]
▽ BASF commercial MeOH synthesis catalyst, prereduced	gas diffusion, 65 atm	0.50	7.5	Satterfield and Cadle [292]
⊽ Girdler G-52 commercial catalyst, 33% Ni on refractory oxide support	gas diffusion, 65 atm	0.44	4.5	Satterfield and Cadle [292]
⧺ Girdler G-58 commercial catalyst, Pd on alumina	gas diffusion, 65 atm	0.39	2.8	Satterfield and Cadle [292]

that only the macropores contributed to diffusion and only the macropore void fraction is used for the correlation in Figure 1.5. Unrealistically high values of τ (e.g., 20 or higher) have been reported when investigators have assumed diffusion to be completely in the bulk range whereas in fact a considerable portion occurred in the transition region (see Section 1.7).

The data points of Ma [297] and Pelossof [298] represent studies of the hydrogenation of α-methylstyrene to cumene in the liquid phase on catalyst pellets of different sizes, from which an effective diffusivity and tortuosity can be calculated by the methods outlined in Chapter 3. Pelossof's catalyst was palladium impregnated on 0.825-cm-diameter alumina spheres. Ma's catalyst, $\frac{1}{8}$-in. by $\frac{1}{8}$-in. alumina pellets, was made by a commercial catalyst manufacturer, the powder being impregnated with palladium before the pellets were formed. Cadle [292] reported studies on counterdiffusion of helium and nitrogen at pressures up to 65 atm through five commercial pelleted catalysts. From the pore-size distribution and the effect of pressure on the helium flux, it appears that at 65 atm diffusion was essentially by the bulk mode. Commercial catalysts are generally calcined at high temperature for mechanical strength, so it is not surprising that their tortuosity factors are somewhat higher than those for pressed powders, which in their more open forms more closely resemble beds of unconsolidated materials. The variation between prereduced methanol synthesis catalysts from three manufacturers (all zinc promoted with chromium oxide) is noteworthy, although the tortuosity factors for the seven commercial catalysts in Figure 1.5 all fall between about 3 and 7.5.

Figure 1.5 shows that there is no unique relationship between τ and θ for porous catalysts. Values of τ of 3 to 7 are readily reconciled with physical reality, however, by multiplying together physically reasonable values of L' and S'. It is perhaps remarkable that the range of values is this small when one considers that the tortuosity factor includes all the deviations between the real porous structure and the idealized model, including dead-end pores, skin effects, and various unknown degrees of anisotropy and inhomogeneity.

In the above analysis, τ is taken to be characteristic of the porous structure, but not of the nature of the diffusing molecules. When the size of the diffusing molecule begins to approach that of the passageway, the wall would be expected to exert a retarding influence on the diffusion flux. As an example, Maxted and Elkins [207] studied the hydrogenation of cyclohexene and of ethyl crotonate in alcoholic solution at 20°C and 1 atm on a platinum catalyst supported on one of several oxides. On an alumina support with average pore radius of 21 Å the rate of hydrogenation of ethyl crotonate was much lower than would have been expected from results on supports of higher pore radius. (See also Section 1.7.4, "Diffusion in Zeolites.")

1.7 Diffusion in Fine Pores

1.7.1 *Knudsen Diffusion*

If the gas density is low, or if the pores are quite small, or both, the molecules collide with the pore wall much more frequently than with each other. This is known as "Knudsen flow" or "Knudsen diffusion." The molecules hitting the wall are momentarily adsorbed and then given off in random directions (diffusively reflected). The gas flux is reduced by the wall "resistance" which causes a delay because of both the diffuse reflection and the finite time the molecule is adsorbed. Knudsen diffusion is not observed in liquids. Kinetic theory provides the following relations for Knudsen diffusion in gases *in a straight round pore*:

$$N = \frac{D_K}{x_0}(c_1 - c_2) = \frac{D_K}{RT}\frac{(p_1 - p_2)}{x_0} = \frac{2r_e\bar{u}}{3RT}\frac{(p_1 - p_2)}{x_0}$$

$$= \frac{2r_e}{3RT}\left(\frac{8RT}{\pi M}\right)^{1/2}\frac{(p_1 - p_2)}{x_0}, \tag{1.30}$$

$$D_K = 9700 r_e \sqrt{T/M}. \tag{1.31}$$

In Equation 1.31 r_e is the pore radius in cm, T the temperature in °K, and M the molecular weight. The symbols refer to a single component; since molecular collisions are negligible, flow and diffusion are synonymous, and each component of a mixture behaves as though it alone were present.

The internal geometries of consolidated porous solids are but poorly understood, and an empirical factor must be introduced to make the theory useful. The straight round pore of radius r_e has a volume-to-surface ratio of $\frac{1}{2}r_e$. The porous material has a total surface S_g (cm^2/g), and an average bulk or pellet density ρ_p (g/cm^3). We may logically define the mean pore radius as

$$r_e = \frac{2V_g}{S_g} = \frac{2\theta}{S_g\rho_p}. \tag{1.32}$$

With this substitution, the Knudsen diffusion coefficient *for a porous solid* becomes

$$D_{K,\text{eff}} = \frac{D_K\theta}{\tau_m} = \frac{8\theta^2}{3\tau_m S_g \rho_p}\sqrt{\frac{2RT}{\pi M}} = 19\,400\,\frac{\theta^2}{\tau_m S_g \rho_p}\sqrt{\frac{T}{M}}. \tag{1.33}$$

As in Equation 1.29, the void fraction θ has been introduced so that the flux N given by $D_{K,\text{eff}}$ will be based on the total cross section of porous solid, not just the pore cross section. As with bulk diffusion, the factor τ_m allows for both the tortuous path and the effect of the varying cross section of individual pores. The subscript m reminds us that τ_m is the value of the tortuosity factor obtained when D_K is calculated from a mean pore radius, Equation 1.32. If a range of pore sizes exists, the proper average pore radius to use in Equation 1.31 is given by Equation 1.50 (see Section 1.8) rather than Equation 1.32, provided that the flux is completely in the Knudsen range.

1.7.2 *The Transition Region*

Bulk diffusion occurs when the collisions of molecules with the pore wall are unimportant compared to molecular collisions in the free space of the pore. Knudsen diffusion occurs when this condition is reversed. In a given pore there is a range of molecular concentrations in which both types of collisions are important. This is the "transition" region. As pressure is reduced the change from bulk to Knudsen diffusion, however, does not occur suddenly when the mean free path of the gas molecules becomes equal to the pore radius. Until fairly recently only empirical or complex expressions of limited usefulness were available for use in the transition region between ordinary and Knudsen diffusion. Almost simultaneously two independent groups of workers [101, 312] derived the same type of relation. For binary gas diffusion in a porous solid at constant total pressure, this takes the form

$$N_1 = \frac{-(P/RT)(dY_1/dx)}{\dfrac{1-(1+N_2/N_1)Y_1}{D_{12,\text{eff}}} + \dfrac{1}{D_{K1,\text{eff}}}}$$

$$= \frac{-D_{\text{eff}}P}{RT}\frac{dY_1}{dx} = -D_{\text{eff}}\frac{dc_1}{dx}, \tag{1.34}$$

where D_{eff} is defined as the ratio of the flux N_1 to the concentration gradient, whatever the transport mechanism. The coefficient $D_{12,\text{eff}}$ is the effective ordinary diffusion coefficient, as given by Equation 1.29, and $D_{K1,\text{eff}}$ is the Knudsen coefficient for constituent 1 as given by Equation 1.33. N_2/N_1 is positive for cocurrent diffusion and negative for countercurrent diffusion.

It is important to be clear on how a flux or a diffusion coefficient is defined, since in some cases the plane of no net molar flux is fixed with respect to fixed coordinates, but in other cases it moves. In studying diffusion in porous substances we generally measure N, *not* J, if the two are different. Furthermore, for most engineering purposes, including diffusion in porous catalysts, we are concerned with fluxes relative to the catalyst, i.e., N. D_{eff} is defined by Equation 1.34 to give the flux relative to the experimental apparatus, *not* relative to the plane of no net molar flux. With nonequimolar counterdiffusion in the transition region or in the bulk-type mode, $N \neq J$ and D_{eff} becomes a function of N_1/N_2. The value of N_1/N_2 in turn depends on the physical situation. For bulk diffusion conditions, $D_{12,eff} \ll D_{K1,eff}$, and Equation 1.34 reduces to

$$D_{eff} = \frac{D_{12,eff}}{1-(1+N_2/N_1)Y_1}. \tag{1.35}$$

Whether Knudsen or ordinary diffusion predominates depends on the ratio D_{12}/D_K and not solely on the pore size or pressure. Note that D_{12} varies inversely with pressure, and does not depend on pore size; D_K is proportional to pore diameter and independent of pressure. If D_{12}/D_K is large, Equation 1.34 reduces to

$$N_1 = -D_{K1,eff}\frac{dc_1}{dx}. \tag{1.36}$$

In the other limit, with D_{12}/D_K small and $N_2 = 0$, the relation becomes

$$N_1 = -\frac{D_{12,eff}P}{p_2}\frac{dc_1}{dx}, \tag{1.37}$$

where p_2 is the partial pressure of constituent 2. For self-diffusion or equimolal counterdiffusion of gases, $N_1 = -N_2$ at constant P, and Equation 1.34 reduces to

$$N_1 = -\frac{1}{(1/D_{12,eff})+(1/D_{K1,eff})}\frac{dc_1}{dx} \tag{1.38}$$

and

$$\frac{1}{D_{eff}} = \frac{1}{D_{K,eff}} + \frac{1}{D_{12,eff}}. \tag{1.39}$$

This result was obtained earlier by Pollard and Present [255]. The form of Equation 1.39 emphasizes that resistance to the motion of molecules of type 1 is caused by collisions with other gaseous molecules or by collisions with the wall or both.

For steady-state binary diffusion at constant P, integration of Equation 1.34 from $(Y_1)_1$ to $(Y_1)_2$ over the length x_0 yields

$$N_1 = +\frac{D_{12,\mathrm{eff}}P}{RTx_0(1+N_2/N_1)} \times \ln\left[\frac{1-(1+N_2/N_1)(Y_1)_2+(D_{12,\mathrm{eff}}/D_{K1,\mathrm{eff}})}{1-(1+N_2/N_1)(Y_1)_1+(D_{12,\mathrm{eff}}/D_{K1,\mathrm{eff}})}\right] = -\frac{D_{\mathrm{eff}}P}{RTx_0}(Y_2-Y_1). \quad (1.40)$$

This reduces to Equation 1.13 when ordinary diffusion predominates ($D_{12,\mathrm{eff}} \ll D_{K1,\mathrm{eff}}$) and to Equation 1.30 for Knudsen diffusion (with $D_{K1,\mathrm{eff}}$ replacing D_K). Equation 1.40 is confirmed by the data of Scott and Dullien [312] on binary diffusion in several porous solids over a 50-fold range in total pressure. The second equality gives a means of defining the effective diffusion coefficient in the transition zone when $N_2/N_1 \neq -1$.

$$D_{\mathrm{eff}} = \frac{D_{12,\mathrm{eff}}}{(1+N_2/N_1)(Y_1-Y_2)} \times \ln\left[\frac{1-(1+N_2/N_1)(Y_1)_2+D_{12,\mathrm{eff}}/D_{K1,\mathrm{eff}}}{1-(1+N_2/N_1)(Y_1)_1+D_{12,\mathrm{eff}}/D_{K1,\mathrm{eff}}}\right]. \quad (1.41)$$

The above expression appears complicated because we are applying Fick's first law to a region in which the diffusion coefficient is not really constant.

If bulk diffusion predominates [i.e., $(D_{12}/D_K) \to 0$], Equation 1.41 reduces to

$$D_{\mathrm{eff}} = \frac{D_{12,\mathrm{eff}}}{(1+N_2/N_1)(Y_1-Y_2)} \ln\left[\frac{1-(1+N_2/N_1)(Y_1)_2}{1-(1+N_2/N_1)(Y_1)_1}\right]. \quad (1.42)$$

Example 1.3 Estimation of D_{eff} for Gas Diffusion in a Porous Catalyst

Estimate D_{eff} for the diffusion of thiophene in hydrogen at 660°K and 30 atm in a catalyst having a B.E.T. surface of 180 m^2/g, a void volume of 40 per cent, a pellet density of 1.40 g/cm^3, and exhibiting a narrow pore-size distribution.

For this system, D_{12} was found to be 0.052 cm²/sec by the procedure illustrated by Example 1.1. Substituting in Equation 1.29, we obtain

$$D_{12,\,\mathrm{eff}} = \frac{D_{12}\,\theta}{\tau} = \frac{0.052 \times 0.40}{\tau} = \frac{0.0208}{\tau}\ \mathrm{cm^2/sec}.$$

Substituting in Equation 1.33, we have

$$D_{K,\,\mathrm{eff}} = \frac{19\,400 \times 0.4^2}{\tau_m \times 1\,800\,000 \times 1.40}\sqrt{\frac{660}{84}} = \frac{0.00344}{\tau_m}\ \mathrm{cm^2/sec}.$$

Taking $\tau_m = 2$, $D_{12,\,\mathrm{eff}} = 0.0104$ and $D_{K1,\,\mathrm{eff}} = 0.00172$ cm²/sec. Knudsen diffusion may be expected to predominate, since $D_{12,\,\mathrm{eff}}$ is so much larger than $D_{K1,\,\mathrm{eff}}$. As a good approximation, Equation 1.39 may be used to give

$$\frac{1}{D_{\mathrm{eff}}} = \frac{1}{D_{K,\,\mathrm{eff}}} + \frac{1}{D_{12,\,\mathrm{eff}}} = 673, \qquad D_{\mathrm{eff}} = 0.0015\ \mathrm{cm^2/sec}.$$

The effect of total pressure on diffusion of gases in pores obviously depends on the relative importance of Knudsen and ordinary, or bulk, diffusion. The manner in which the diffusion flux varies with total pressure is illustrated in Figure 1.6. The curve shown is calculated for

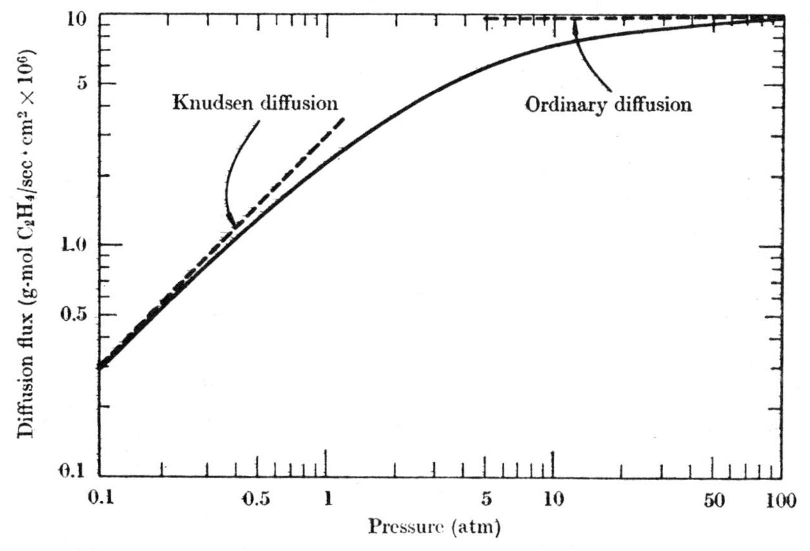

Figure 1.6. Steady-state counterdiffusion of ethylene and hydrogen in the transition region calculated by the approximate Equation 1.38 for a porous plug 1.0 cm thick, with pure ethylene at P on one side and pure hydrogen at P on the other. $S_g = 10$ m²/g; $T = 298$°K; $\tau = 1.0$; $\theta = 0.4$; $\rho_p = 1.4$ g/cm³; $D_{12}P = 0.602$ cm²/sec; $r_e = 570$ Å.

the case of a porous disk exposed to pure ethylene and pure hydrogen at the two faces, using the approximate Equation 1.38. Knudsen diffusion predominates at low pressure, and the flux increases with pressure because the concentration difference across the disk increases (D_K is independent of pressure). At high pressure, the flux rate approaches the constant value for ordinary diffusion; this is constant since the concentration difference is directly, and D_{12} inversely, proportional to pressure.

Figure 1.7 illustrates the effect of pore size at fixed pressure on the

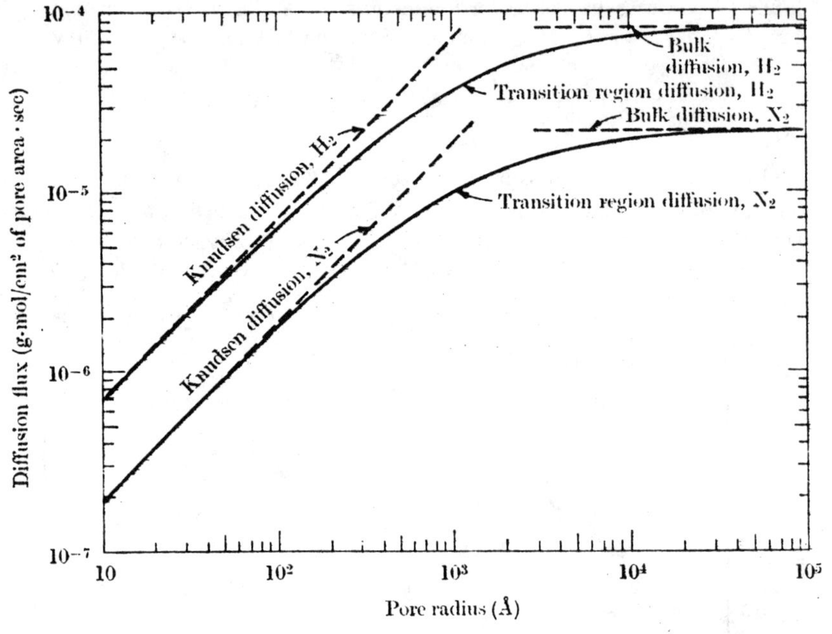

Figure 1.7. Counterdiffusing flux, N, of hydrogen and nitrogen as a function of pore size. Total pressure, 1 atm, $T = 298°K$. Saraf [290].

diffusion flux N for the binary system hydrogen-nitrogen. The ratio of the fluxes of the two species is inversely proportional to the square root of the ratio of their molecular weights not only in the Knudsen and transition regime, but also into the bulk regime when the pores are of the order of a micron in size. As the pores become of the order of 10 μ or larger, the counterdiffusion molar fluxes approach equality [311, 312]. The range of pore sizes for which diffusion will occur in the transition region depends on the relative values of D_K and D_{12} and moves in the direction of smaller pores with increased pressure. To illustrate the effect of composition, Table 1.7 gives the transition region limits for

Table 1.7 Transition Region[a] for Selected Binary Gas Pairs; 300°C, 1 atm

	Limiting Pore Radius (Å)	
	Lower	Upper
H_2-CO	245	26 300
H_2-C_6H_6	137	18 100
H_2-$C_{15}H_{30}$ (pentadecene)	19	3 180
$C_{15}H_{30}$-$C_{15}H_{32}$ (pentadecane)	8	535
air-$C_{10}H_8$ (naphthalene)	76	6 350

[a] Diffusion flux 10 per cent or more below that predicted for Knudsen or bulk diffusion, respectively. Cadle [53].

some binary gas pairs at 300°C and 1 atm, arbitrarily defining the transition region as that in which the diffusion flux is 10 per cent or more below that predicted by the Knudsen or bulk diffusion equations, respectively [53].

The effect of temperature is also different for the two diffusion mechanisms. As noted earlier, D_{12} for gases and liquids increases moderately with temperature, corresponding activation energies being in the range 2 to 5 kcal/g-mol. The Knudsen diffusion coefficient, however, increases as the square root of the temperature. The corresponding values of the activation energy are 0.3 kcal at 20°C and 0.77 kcal at 500°C (when expressed in g-cal, the values are approximately equal to the absolute temperature).

1.7.3 *Surface Diffusion*

Molecules adsorbed on solid surfaces may evidence considerable mobility. Transport by movement of molecules over a surface is known as "surface diffusion" and the direction is that of decreasing surface concentration. Equilibrium adsorption is a function of the partial pressure of the adsorbed constituent in the gas adjacent to the surface, and both tend to decrease in the direction of diffusion. Thus, surface and gas-phase diffusion proceed in parallel. Surface diffusion has been reviewed recently by Barrer [18], by Dacey [84], and by Field, Watts, and Weller [105]. Evidently, it cannot be significant unless appreciable adsorption occurs, yet if adsorbed molecules are held so strongly as to be essentially immobile, surface diffusion will be insignificant.

In steady-state counterdiffusion through a finely porous pellet, the flux ratio N_2/N_1 should be proportional to $\sqrt{M_1/M_2}$. If measurements are made with helium and a second gas as a counterdiffusing pair, any excess flux of the second gas over that calculated by this relationship of flux to molecular weight is attributed to surface diffusion. Surface diffusion of helium has been believed to be insignificant at ambient temperature and pressure [18, 105, 360], since it is adsorbed so slightly, but a small amount of surface diffusion on porous Vycor has been reported by Hwang and Kammermeyer [165].

An alternate procedure, applicable only in the completely Knudsen region, is to measure the forced-flow flux of helium and of another gas singly through a particular porous solid. Again, any excess flux over the theoretical is attributed to surface diffusion. There is considerable controversy over how best to describe surface mass transport, but especially for amounts of adsorption corresponding to a fraction of a monolayer, which occurs when the partial pressure of the gas is substantially below the vapor pressure (e.g., 0.05 or less) a Fick's law type relationship is usually applied. The proportionality constant D_s varies somewhat with surface concentration, generally being greater at higher values of the surface concentration, but approaching some limiting value as surface coverage is decreased. The surface diffusion flux per unit cross section area of porous catalyst J_s may thus be expressed by

$$J_s = \frac{-D_s}{\tau_s} \rho_p S_g \frac{dc_s}{dx}, \tag{1.43}$$

where

$$D_s = D_0 e^{-E_s/RT} \tag{1.44}$$

and τ_s accounts for the tortuous path of surface diffusion, c_s is the surface concentration (mol/cm^2), D_s is in cm^2 of pellet cross section per second, and the product $\rho_p S_g = S_v$ is the surface area per unit volume (cm^2/cm^3 of pellet volume). There is no reason why τ_s should equal τ as used previously, although this assumption is sometimes made in the absence of any other information. Surface diffusion will most likely be significant with high area and, therefore, fine-pore pellets. If it is assumed that the bulk gas diffusion is all by the Knudsen mode, the ratio of surface to Knudsen flux is then

$$\frac{J_s}{J_K} = \frac{\tau_K D_s}{\tau_s D_K} \frac{\rho_p S_g}{\theta} \left(\frac{dc_s}{dc_g}\right). \tag{1.45}$$

At low surface coverages the adsorption is closely approximated by Henry's law and $dc_s/dc_g = K_H$, the absorption equilibrium constant (cm^3 of gas/cm^2 of surface).

Under unsteady-state conditions, in which diffusion occurs simultaneously with adsorption, the following relation holds for flat-plate geometry:

$$\frac{\partial c_g}{\partial t} = \left[\frac{D_{K,\text{eff}} + K_H D_{s,\text{eff}} \rho_p S_g}{\theta + K_H \rho_p S_g}\right] \frac{\partial^2 c_g}{\partial x^2}, \tag{1.46}$$

where $D_{s,\text{eff}} = D_s/\tau_s$.

This assumes that adsorption equilibrium is established at all times and all positions. If adsorption is unimportant Equation 1.46 reduces to

$$\frac{\partial c_g}{\partial t} = \left(\frac{D_K}{\tau_K}\right) \frac{\partial^2 c_g}{\partial x^2}, \tag{1.47}$$

where D_K is based on the void portion of the pellet cross section. Equation 1.46 has important implications for interpretation of diffusion measurements obtained by unsteady-state methods such as gas chromatography. Transient methods give the quantity in brackets in Equation 1.46 which may be either greater or less than D_K/τ_K. The term $K_H D_{s,\text{eff}} \rho_p S_g$ reflects the contribution of surface diffusion to the flux and the term $K_H \rho_p S_g$ the effect on the flux caused by accumulation (or depletion) of adsorbed material. *Even when surface mobility is unimportant*, accumulation of material by adsorption may be of major significance, but this is sometimes overlooked in analysis of data. The term θ will always be less than unity but $K_H \rho_p S_g$ can greatly exceed unity with high-area catalysts (e.g. 100 m^2/g or higher) and adsorption corresponding to a substantial fraction of a monolayer (e.g., 0.1 or more).

Values of D_s at low surface coverages are typically in the range of 10^{-3} to 10^{-5} cm^2/sec at ambient temperature for physically adsorbed molecules such as hydrogen, nitrogen, krypton, carbon dioxide, methane, ethane, propane, and butane on such surfaces as porous glass, carbon, silica gel, and typical commercial catalysts supported on porous oxides, such as alumina [18, 119, 293, 308, 323]. The activation energy E_s in most cases is approximately one-half of the energy of physical adsorption, but in a few cases closely approaches it in value [308]. J_s should then always decrease with increasing temperature since K_H and thus c_s will decrease more rapidly than D_s will increase.

Sladek [323] has recently developed the general correlation of surface diffusion coefficients shown in Figure 1.8, which brings together data on a wide variety of systems, both of physical adsorption and chemisorption. Since surface mobility should be related to strength of adsorption, D_s is plotted against the parameter q/mRT, where q is the heat of adsorption and m is an integer having a value of 1, 2, or 3 depending upon the nature of the surface bond and of the solid, as shown in Table 1.8.

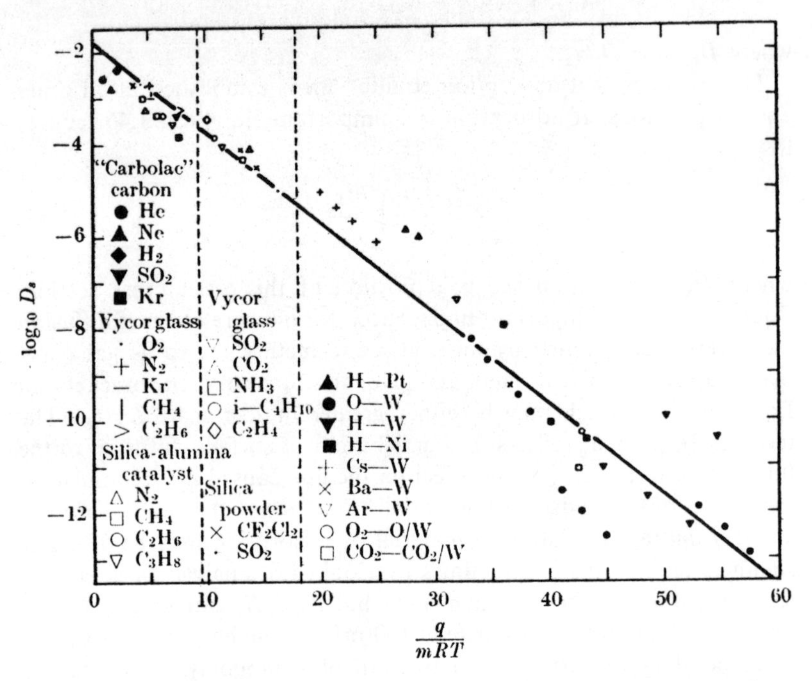

Figure 1.8. Correlation of surface diffusion coefficients. Sladek [323].

For hydrogen and oxygen on metals, adsorption with dissociation was assumed; q was taken to be equal to one-half of the measured molecular heat of adsorption plus one-half of the dissociation energy of the diatomic gas. The assignment of values of m was made to achieve the best correlation of the data, and represents the ease with which an adsorbed molecule can migrate relative to the strength of bonding; a larger value of m corresponds to a larger value of D_s, for a fixed value of q/RT.

Certain symbols are used more than once in Figure 1.8; three sets of

Table 1.8 Values of m for Use in Figure 1.8

Bond	Solid	m	Examples of Available Surface Diffusion Data
Van der Waals			
Polar Adsorbate	Conductor	2	SO_2-Carbon
	Insulator	1	SO_2, NH_3-Glass
Nonpolar Adsorbate	Conductor	1	Ar-W, N_2-Carbon
	Insulator	1	Kr, C_2H_4-Glass
Ionic	Conductor	2	Cs, Ba-W
	Insulator	1	none
Covalent	Conductor	3	H-Metals, O-W
	Insulator	1	none

legends are shown, each applying to a range of values of q/mRT. The data points shown for physically adsorbed species are a representative selection from the much greater quantity of data used in the original correlation. The correlation is remarkable in that it represents eleven orders of magnitude of D_s to within about one and a half orders of magnitude, and over a q/mRT range of 60. Values of q from 0.3 to 200 and temperatures of -230 to $+600$°C are represented as well as data obtained by a variety of systems and experimental techniques. Kammermeyer and Rutz [166] have published a useful correlation shown in Figure 1.9, which also provides a simple visualization of the contribution of surface diffusion to the total flux for a number of gases and vapors. The product of the effective diffusivity at 25°C and the square root of the molecular weight, $D_{1,\text{eff}}\sqrt{M_1}$, is plotted as a function of either the boiling point or the critical temperature of the diffusing species. Taking the contribution of surface diffusion to the flux as being negligible for helium, the difference between the curve and the flat dashed line represents the contribution of the surface flux for other species. These data are for diffusion through porous Vycor of 200 m^2/g adjusted in some cases where different samples were used and were obtained at upstream pressures below 3 atm, downstream pressure of 1 atm, under which conditions the gas phase flux was completely in the Knudsen region. On higher area substances surface diffusion will make a larger contribution. For example, on "Carbolac" carbon plugs with an average pore radius of 19 Å, the surface diffusion flux of nitrogen, argon, or krypton at ambient conditions was from a third to the equal of the Knudsen flux [18]. Schneider and Smith [308] used a chromatographic method to measure surface diffusion coefficients for ethane, propane, and n-butane on silica gel (832 m^2/g; $r_a = 11$ Å) over the temperature range of 50–175°C and reviewed all other available information on surface diffusivities of these hydrocarbons on various substrates. The activation energies for the hydrocarbons were usually from about 3 to 5 kcal. At 50°C surface diffusion accounted for 64, 73, and 87 per cent of the transport for ethane, propane, and butane, respectively. At 125°C, surface transport of n-butane represented 68 per cent of the total.

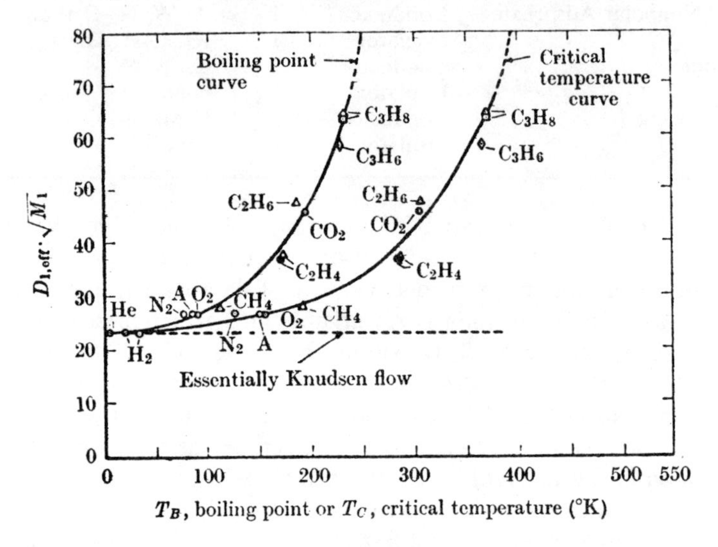

Figure 1.9. Correlation of the product $(D_{1,\text{eff}})(M_1)^{1/2}$ with boiling point T_B and critical temperature T_C for a series of gases through porous Vycor glass of 200 m^2/g. Measurements at 25°C and 1 to 3 atm. Kammermeyer and Rutz [166].

Studies by Cadle [293] of counterdiffusing helium and nitrogen through five commercially pelleted catalysts (three methanol synthesis catalysts and two metal catalysts supported on oxides) at ambient temperatures showed that the surface diffusion of nitrogen made an increasing contribution to the total nitrogen flux with increase in pressure level. At the highest pressure, 65 atm, the surface diffusion flux was of comparable magnitude to the volume diffusion flux. The adsorption equilibrium constant was independent of pressure but the gas-phase mode of diffusion varied from predominantly Knudsen at atmospheric pressure to predominantly bulk at 65 atm. Equation 1.45 shows that the ratio of surface diffusion flux to gas-phase diffusion flux should indeed increase with pressure level, since the effective gas-phase diffusion coefficient will begin to drop with increasing pressure level. With the wide pore-size distribution typical of most commercial catalysts, rigorous analysis of the pressure effect is complex since the relative contributions to the total flux from pores of different sizes varies with pressure level.

With chemisorbed species, D_s is relatively small and the surface approaches saturation at pressures that may be several orders of magnitude less than occurs for physical adsorption of the same gaseous species at the same temperature. Consequently even on the highest surface area porous metal compacts available, surface diffusion is a significant portion of the total flux only at very low pressures (e.g., 40 μ Hg for hydrogen on nickel, below about 1 mm Hg for hydrogen on platinum), where low surface coverage permits a steep slope of the adsorption isotherm (dc_s/dc_g) in Equation 1.45.

The above discussion suggests that surface diffusion will make no significant contribution to intraparticle flux in porous catalysts at highly elevated temperatures, although there is little direct evidence one way or the other. When the partial pressure of the reacting species is an appreciable fraction of the vapor pressure, surface diffusion may be of considerable importance, as has been suggested to explain studies of the catalytic dehydration of ethanol at 400–700°F and pressures up to 100 psi on a silica-alumina bead catalyst [177, 217]. Since gaseous diffusivities in porous substances are usually measured at ambient or only slightly elevated temperatures, results of such studies, particularly on high-area materials, must be analyzed most carefully to be sure that the possible contribution from surface flux has been isolated and taken into account.

Somewhat more conclusive evidence for significant surface diffusion in a chemical reaction is presented by Bienert and Gelbin [29] who studied the dehydration of isopropanol at 200–250°C on γ-alumina in the form of either powder or pellets compressed from the powder to various densities. By comparison of the rates on the two different sizes, effectiveness factors of the various pellets were found in the range of 0.088–0.71. Back-calculation (see Chapter 3) gave values of the effective diffusivity varying from 0.015–0.06 cm^2/sec, which for a given pellet reached a maximum value at a temperature intermediate between the extremes of 200–250°C studied. A study by Sterrett and Brown [329] of the *o-p* hydrogen conversion at 32 psia and 76°K gave values of the effective diffusivity back-calculated from reaction rate data substantially higher than would be expected, indicating that surface diffusion may have been significant at this low reaction temperature, a conclusion that seems plausible since the catalyst, a ferric oxide gel, contained pores that were all below about 30 Å radius.

Analysis of surface diffusion effects generally assumes that the thickness of the adsorbed layer of gas is not an appreciable fraction of the pore cross section, i.e., that the adsorbed layer does not affect the effective pore size for Knudsen diffusion. Evidently, this must become invalid for sufficiently small pores, and cases in which this effect seems to be significant have been cited [91, 106, 292, 360].

As the size of a molecule becomes an appreciable fraction of the pore diameter, the molecule comes under the influence of the wall during most of its movement and representation of any of the flux as Knudsen-type diffusion must gradually become invalid. Little is known about conditions under which this occurs or the nature of diffusion under these conditions. (See Section 1.7.4 below and Section 3.5.)

1.7.4 *Diffusion and Reaction in Zeolites* ("*Molecular Sieves*")

The physical structure of zeolites is summarized in Section 1.5.2. Sorption and diffusion characteristics have been recently reviewed by Walker and co-workers [357]. Most diffusion studies have been on the small-pore zeolites (4–5 Å diameter pores) in which, for nonpolar gases comprising small molecules such as nitrogen, methane, propane and *n*-butane, diffusion coefficients are in the range of 10^{-12}–10^{-14} cm^2/sec at temperatures of 20–200°C and the apparent activation energies are in the range of 3 to 11 kcal/mole. These diffusivities are very small compared to typical values of 10^{-3} to 10^{-4} cm^2/sec found for the same gases in such high-area catalysts as conventional silica-alumina. In sodium mordenite (6 × 7 Å diameter pores) diffusion coefficients at 25°C varied from about 33×10^{-10} cm^2/sec for methane to 12×10^{-10} cm^2/sec for n-C_4H_{10} [295]. From cracking studies at 300–540°C with *n*-hexane on H-offretite (pores 3.7 by 4.1 Å) which were apparently highly diffusion limited, Miale, Chen, and Weisz [214] concluded that the effective diffusivity for this system was less than 10^{-9} cm^2/sec. Little information is available on diffusion rates in the zeolites having larger pore sizes. From counterdiffusion studies of benzene and cumene at 25°C on catalyst Linde SK-500, a rare-earth exchanged form of Y zeolite (about 10 Å minimum diameter in the Na form), Katzer [168] concluded that diffusion coefficients were not less than 10^{-11} cm^2/sec, but they may have been much larger. On NaY, diffusion coefficients for cumene diffusing out into benzene were 9×10^{-14} and 1.5×10^{-12} at 25° and 65°, respectively. On H-mordenite equilibrium could not be obtained, but diffusivities of cumene under desorption conditions were about 10^{-13} (65°C) to 10^{-14} (25°C) cm^2/sec.

Evidently as molecular size and bulkiness of the reactant molecule is increased, a point will be reached at which diffusion of the reacting molecules into the pores of a particular zeolite will represent a limitation on the observed rate of reaction, but little is known about conditions under which this occurs. Weisz and co-workers [373] compared the reaction rates of various compounds in binary mixtures of the straight chain and branched isomer forms, on the small-pore 4A and 5A zeolites and also on 10X. As examples of their results, on 5A zeolites at 500°C (5 Å diameter pores), cracking of 3-methyl pentane was essentially prevented by diffusion limitation, whereas that of the *n*-hexene present in the mixture was not, but no diffusion limitations were exhibited by either compound on the 10X sieve (about 8 Å pore diameter). 1-butanol was selectively dehydrated on the 5A sieve but not the 2-methyl-1-propanol present. Catalysts with noble metals incorporated inside the cavities were shown to hydrogenate selectively *n*-olefins in the presence of *iso*-olefins or to oxidize *n*-paraffins or *n*-olefins in the presence of the branched isomers. Further applications of shape-selective catalysts have been described by Chen and Weisz [67]. On 5A sieve *trans*-butene-2 is hydrogenated from three to seven times faster than *cis*-butene-2, in mixtures of the two; the diffusion of the *trans*-isomer into the zeolite was also shown to be substantially greater than that of the *cis* isomer. On a sodium mordenite zeolite into which platinum had been incorporated, ethylene was selectively hydrogenated to ethane in the presence of propylene, at temperatures of 175–260°C, e.g., in a 1:1 mole ratio in the presence of hydrogen at 260°C. A 28 per cent conversion of ethylene was obtained with less than 0.1 per cent conversion of propylene. Shape-selective catalysts may also be used as an internal method of generating or removing heat in a reactor. In a mixture of 2 per cent CO and 0.5 per cent *n*-butane in air at 427–540°C a platinum-loaded zeolite A was shown to cause 86 to 100 per cent reaction of the carbon monoxide and only 2.4 to 5.5 per cent reaction of the butane.

Diffusion limitations can also develop by the buildup in the cavities of reaction products of higher molecular weight than that of reactants, as demonstrated for alkylation reactions by Venuto and Hamilton [347]. Norton [228] showed that for propylene polymerization, activity decreased in the order 10X $>$ 13X $>$ 5A $>$ 4A $>$ 3A, that of 3A being zero. The data were interpreted in terms of differing intrinsic catalyst activities, but the results may have been markedly affected by the relative rates of diffusion of polymer product out of the pores.

Bryant and Kranich [49] studied the dehydration of ethanol and *n*-butanol over types X, A, and synthetic mordenite after the zeolites had been base-exchanged with various cations. The reaction rate and ratio of olefin to ether in the product were markedly affected by the type of molecular sieve and by the size and charge of the exchanged cation.

Much more remains to be learned about diffusion in zeolites. The flux may not follow Fick's law closely and diffusion coefficients can be markedly affected by slight blockage of passageways. Furthermore, although pore structure is usually regarded as perfect and regular for purposes of analysis, zeolites are unquestionably polycrystalline and present a heterogeneous set of passageways. In some cases the reaction may occur in part or exclusively on sites outside the fine intracrystalline pores.

1.8 Estimation of Diffusion Coefficients in Porous Solids

Table 1.9 and Figure 1.10 present the results of thirteen studies of diffusion or flow which illustrate some of the factors involved in predicting the effective diffusion coefficient. Table 1.10 presents additional data not shown in the figure. Since Knudsen flow predominated in all of the tests represented, flow and diffusion should be synonymous. The bulk-diffusion term has been ignored and $D_{eff} \times 10^4$ plotted as ordinate with the group $[\theta^2(T/M)^{1/2}/S_g \rho_p] \times 10^8$ as abscissa. The lines shown have been drawn with slopes of unity; as read from the figure, an intercept with the ordinate (at abscissa of unity) corresponds to $19400 \times 10^{-4}/\tau_m$.

Porous Vycor glass has been investigated the most frequently by far, at least nine studies being reported. For six, the results agree closely, the extreme values of τ_m being only 4.7 and 6.6 with an average value of 5.9. Three other studies report values of 3.2, 3.3 [141, 360] and 10.5 [310]. For the preparation of porous Vycor a borosilicate glass is formed into the ultimately desired shape and is then annealed at a suitable temperature which allows the formation of two phases, but below that at which physical deformation occurs. One phase is rich in silica and the other rich in sodium oxide and borate. After cooling, the object is leached with dilute hydrochloric acid to leave behind a porous material comprising 96 or more per cent silica. Both phases must be continuous, but the structure of the two phases, and hence that of the resulting pore structure after leaching, may vary substantially with the cooling procedure [275]. It has also been suggested that the thickness

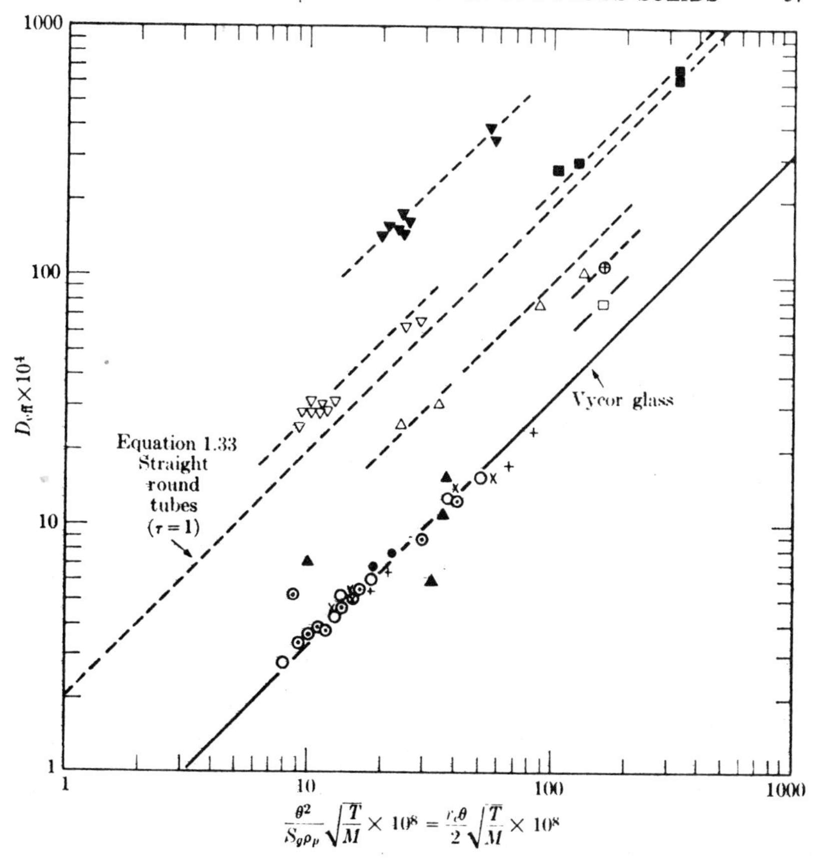

Figure 1.10. Data on diffusion of gases in consolidated porous media compared with equation for Knudsen diffusion in straight round pores. For key, see Table 1.9.

of the initial piece of Vycor may have a significant effect on its diffusion characteristics since this may affect the cooling rate [141]. It is also evident that the leaching procedure may produce a somewhat different porous structure for thin than for thick pieces of glass, since removal of material from the interior of a shape by solution becomes more difficult in thicker geometries.

Three of the studies in Table 1.9 were on silica-alumina commercial bead cracking catalysts. In two of these the effectiveness factor of whole beads was determined from reaction studies by crushing the catalyst and comparing the rate of reaction on whole beads with that on one or more sizes of the crushed catalyst. By the methods outlined in Chapter 3,

Table 1.9 Diffusion and Flow in Finely Porous Media (Experimental data shown in Figure 1.10)

Material	Technique	Gases	T(°K)	r_e (Å)	θ	τ_m[a]	Reference
■ Alumina plug compressed from powder	Diffusion	N_2, He, CO_2	303	96	0.812	0.85	Henry, Chennakesawan, and Smith [140]
△ Fresh and regenerated silica-alumina commercial bead cracking catalyst	Flow	H_2, N_2	298	31–50	0.464	2.1	Villet and Wilhelm [348]
× Vycor glass	Flow	H_2, He, Ar, N_2	298	30.6	0.31	5.9[b]	Gilliland, Baddour, and Russell [119]
⊕ Water-gas shift catalyst	Diffusion	O_2, N_2	298	177	0.52	2.7	Bokhoven and van Raayen [39]
□ Ammonia synthesis catalyst	Diffusion	O_2, N_2	298	203	0.52	3.8	Bokhoven and van Raayen [39]
○ Vycor glass	Flow	He, Ne, H_2, Ar, N_2, O_2, Kr, CH_4, C_2H_6	292, 294	30	0.298	5.9[b]	Barrer and Barrie [19]
+ Vycor glass	Flow	H_2, He, Ar, N_2	298	30	0.28	5.9[b]	Gilliland, Baddour, and Engel [118]

▲ Silica-alumina commercial bead cracking catalysts	Reaction	Gas oil cracking	755	28	—	3	Johnson, Kreger, and Erickson [159]
Catalysts prepared from bead hydrogels	Reaction	Gas oil cracking	755	51–71		3–10	Johnson, Kreger, and Erickson [159]
● Vycor glass	Flow	Ar, N_2	298	46	0.305	5.9[b]	Whang [381]
▽ Silica-alumina cracking catalyst, plug compressed from powder	Flow	He, Ne, Ar, N_2	273–323	16	0.40	0.725	Barrer and Gabor [20]
▼ Silica-alumina cracking catalyst, plug compressed from powder	Flow	He, Ne, N_2	273–323	24	0.53	0.285	Barrer and Gabor [20]
⊙ Vycor glass	Flow	He, CO_2, N_2, O_2, Ar	298	44	0.31	5.9[b]	Kammermeyer and Rutz [166]
⊗ Silica-alumina commercial bead cracking catalyst	Reaction	Cumene cracking	693	(24)	(0.50)	5.6	Weisz and Prater [377]
	Diffusion	H_2, N_2	293	(24)	(0.50)	2.3	

[a] τ_m is based on mean pore radius, Equations 1.32 and 1.33.
[b] Average of five sets of data.

Table 1.10 Diffusion and Flow in Finely Porous Media (Data not shown in Figure 1.10)

Material	Technique	Gases	T(°K)	r_e (Å)	θ	τ_m[a]	Reference
Silica-gel	diffusion (chromatographic technique)	Ethane	323–473	11	0.486	3.35	Schneider and Smith [308]
Houdry-Type A chromia-alumina dehydrogenation catalyst	reaction (carbon burnoff)	O_2	843–915	140(?)	0.56	1.33	Mickley, Nestor, and Gould [216]
Porous Vycor	diffusion	He, Ar	298, 323	20	0.28	3.14	Weaver and Metzner [360]
Porous Vycor	diffusion	He, N_2	299	52.6	0.30	3.3	Henry, Cunningham, and Geankoplis [141]
Porous Vycor	diffusion	H_2, N_2, Ne	298	45	0.30	5.9	Rao and Smith [260]
Porous Vycor	flow	air	298	27.5	0.30	10.5	Schwertz [310]
Catalyst gel spheres	diffusion (unsteady state)	CH_4	295				Gorring and de Rosset [122]
1. 96% silica–4% alumina				15	0.46	5.7	
2. 100% alumina				43	0.68	2.1	
3. 100% silica				28	0.56	3.4	
4. 92% silica–8% alumina				41	0.57	2.5	
5. 90% silica–10% alumina				33	0.51	4.6	

[a] τ_m is based on mean pore radius, Equations 1.32 and 1.33.

it is possible to calculate an effective diffusivity and tortuosity from such data. In both cases it was assumed that the reaction was first order, which introduces an uncertainty into the final result. Nevertheless, the values of τ_m of 3 and 5.6 on commercial bead catalysts compare reasonably well with 2.1 obtained from flow studies [348] and values of 2.1 to 2.5 from diffusion measurements on the same type of catalyst, before and after extensive commercial cracking operations (Table 1.11). Both Vycor and silica-alumina bead catalysts have a fairly narrow pore-size distribution range, the pores in Vycor being essentially all within the range of 10 to 90 Å radius [260]. Three of the studies are on plugs compressed from alumina or silica-alumina powder. In these cases, $\tau_m < 1$, which has no physical meaning and illustrates the inadequacy of using an average pore size defined by Equation 1.32 for prediction of a diffusion coefficient if a substantial pore-size distribution exists. The same conclusion is brought out by studies on commercial catalysts and supports summarized in Table 1.13. For 12 of these on which suitable data are available, τ_m based on an average pore size varied by more than an order of magnitude, from 0.3 to 4. Since D_K is proportional to pore diameter the diffusion flux per unit cross section of pore is much greater for large pores than for small pores, but values of τ_m exceeding unity are found even with catalysts having a wide pore-size distribution. These particular catalysts generally, but not always, likewise exhibit a relatively high value of τ_p as calculated by the parallel-path pore model (see later).

The study of Weisz and Schwartz [378], summarized in Table 1.11, reports data on the diffusion of hydrogen through nitrogen at room temperature and atmospheric pressure in 59 different samples of porous oxide catalysts derived from gels. Most were silica-aluminas though a few samples of silica-magnesia were also tested. The samples differed mainly as to the nature of the thermal or hydrothermal treatment of the gels. Samples prepared by incorporating hard powders in the gel during formation presumably result in wide pore-size distributions although these were not measured. Several samples of used commercial cracking catalyst were also tested. Average pore sizes ranged from 9–250 Å (radius); Knudsen diffusion predominated in all cases.

As with the results in Tables 1.9 and 1.13, τ_m may be either greater than or less than unity and varies by a factor of 20 between extreme cases. With the homogeneous beads (groups A, B, and C) where a narrow pore-size distribution presumably exists, the tortuosity factor τ_m calculated by Equation 1.32 increases with decreasing void fraction,

Table 1.11 Diffusion in Porous Oxide Gel-Derived Catalysts[a]

Catalyst	Pore Volume[b] V_g (cm^3/g)	Particle Density[b] ρ_p (g/cm^3)	S_g (m^2/g)	τ_m[c]	$D_{eff} \times 10^3$ (cm^2/sec)[d]
A. Silica-Alumina "homogeneous" beads					
1. High-density beads, 3 samples	0.21–0.26	1.44–1.56	22–381	7.4–8.7	17.5–0.6
2. Low-density beads, 35 samples	0.32–0.66	0.92–1.3	46–450	1.61–4.7	42–3.0
3. After thermal deactivation for 16 hours at:					
540°C	0.412	1.07	366	2.1	5.5
705	0.381	1.12	269	2.35	6.0
815	0.317	1.20	153	2.6	7.0
870	0.250	1.28	98	3.8	4.9
925	0.121	1.59	29	9.9	1.9
4. From commercial cracking units, 4 samples	—	1.22–1.32	174–272	2.1–2.5	9.6–6.1

B. Chromia-alumina and chromia-alumina-molybdena beads, 2 samples	0.310–0.368	1.60–1.71	155–162	1.56–1.7	20–15
C. Silica-magnesia beads, 1 sample	0.201	1.71	442	8.1	0.45
D. Commercial pelleted clay cracking catalysts, in regenerated state after commercial use, 2 samples	0.26–0.33	0.9–1.01	65–105	0.52–0.83	42–29
E. Pulverized dried silica-alumina dispersed in silica-alumina gel, 1 sample	0.61	0.96	376	0.44	49
F. Same, dispersed in silica-magnesia gel, 2 samples	—	1.22–1.28	348–472	0.62–0.63	19–17

[a] Weisz and Schwartz [378].
[b] Values of ρ_p and V_g by mercury-helium method. For silica-alumina $\rho_t = 2.30$; for silica-magnesia $\rho_t = 2.50$.
[c] τ_m is based on mean pore radius, Equations 1.32 and 1.33.
[d] D_{eff} is based on the diffusion flux of hydrogen through nitrogen at room temperature and atmospheric pressure. The data in the original publication have a typographical decimal-point error.

brought about either by the method of initial preparation or by subjection to high temperature, which causes partial sintering.

From all the above results it follows that, if the pore-size distribution is narrow and the pores of the catalyst are so fine that Knudsen diffusion occurs exclusively, catalysts not subjected to excessive sintering or other treatments markedly affecting the pore structure may be characterized by a value of D_K calculated from Equations 1.31 and 1.32. D_{eff} is calculated by Equation 1.29, replacing D_{12} with D_K and applying a value of τ_m in the reasonable range of 2 to 6. Examples of this type of structure are most silica gels, porous Vycor glass (typical pore radius maxima at about 15 and 40 Å radius, respectively), and silica-alumina bead catalysts [259, 260]. If the pore-size distribution is wide, a different approach is required for prediction (see below).

The values of D_{eff} in Table 1.11 include those for several commercial catalysts and also illustrate the degree of variation in effective diffusivity obtainable by various methods of preparation and subsequent physical treatment. Table 1.12 lists values of D_{eff} for several additional commercial catalysts, which, however, are not as well characterized physically. Table 1.13 (discussed below) lists values of D_{eff} for 17 well-characterized commercial pelleted catalysts.

1.8.1 *Parallel-Path Pore Model*

The most difficult situation to handle occurs with catalysts of wide pore-size distribution, particularly when the pores do not divide clearly into a group of micropores and a group of macropores. The application of elegant models of the pore structure implies a degree of knowledge of the fine structure of a catalyst which, in most cases, does not exist. Indeed, attempts to go beyond the simplest mathematical description of the porous structure, when combined with the nonlinearity of the transition region diffusion equation, lead to models of such complexity that they become difficult either to test or to use. A model related to one proposed by Johnson and Stewart [161] is reasonably close to reality yet relatively simple to use. The pore structure is visualized as comprising an array of parallel cylindrical pores having the pore-size distribution found experimentally. The tortuosity factor (designated as τ_p for this model) is assumed to be independent of pore size or diffusion mode and, as previously, is the adjustable parameter that allows for such factors as varying pore alignment and cross-sectional area. It is assumed that at any cross section perpendicular to the direction of mass transport, the concentration is uniform and independent of pore size, i.e., sufficient cross passages exist physically to make this possible. The

Table 1.12 Diffusion in Selected Commercial Catalysts

	S_g (m^2/g)	$D_{eff}{}^a \times 10^3$	Ref.
1. Silica-alumina, "homogeneous" beads. Laboratory preparations, 3 samples	270–400	6–9	Weisz and Goodwin [374]
2. Silica-alumina homogeneous beads, containing 0.15 wt% Cr, after extensive use in refinery cracking units, 3 samples		6–10	Weisz and Goodwin [374]
3. Similar to 2, but after some thermal damage in a commercial cracking plant, 3 samples		3–4.5	Weisz and Goodwin [374]
4. Pulverized dried silica-alumina dispersed in silica-alumina gel, 1 sample		70	Weisz and Goodwin [374]
5. Chromia-alumina beads, 2 samples		20	Weisz and Goodwin [374]
6. Commercial molecular sieve bead catalyst containing 4.5 or 10% zeolite X exchanged with rare earths, 0.2% Cr_2O_3, and 30–40% inert α-alumina, all dispersed in silica-alumina matrix, average of 15 particles			Cramer, Houser, and Jagel [78]
A. fresh catalyst		27–28	
B. after 3 hr at 1300°F in air		15	
C. fresh catalyst after 24 hr at 1200°F in 15 psig of steam	101–121	28–29	
D. after 5 hr at 1200°F in 100 psig of steam	57–73	36–40	

[a] Diffusion flux of hydrogen through nitrogen at room temperature and atmospheric pressure, divided by concentration gradient (cm^2/sec).

Table 1.13 Tortuosity Factors for Selected Commercial Catalysts and Supports[a]

Designation	Nominal Size	Surface Area (m^2/g)	Total Void Fraction	$D^b_{eff} \times 10^3$ (cm^2/sec)	Average Tortuosity Factor τ_p Parallel-Path Pore Model[c]	$r_e = 2V_g/S_g$ (Å)	τ_m Based on Average Pore Radius[d]
T-126	3/16 × 1/8 in.	197	0.384	29.3	3.7 ± 0.2	29	0.45
T-1258		302	0.478	33.1	3.8 ± 0.2	23.6	0.41
T-826		232	0.389	37.7	3.9 ± 0.1	21.4	0.26
T-314		142	0.488	20.0	7.1 ± 0.9	41.5	1.2
T-310		154	0.410	16.6	3.8 + 0.1	34.3	0.67
G-39	3/16 × 3/16 in.	190	0.354	17.5	4.8 ± 0.3	22.4	0.53
G-35		—	0.354	18.2	4.9 ± 0.1		—
T-606		—	0.115	27.7	2.9 ± 0.2		—
G-58		6.4	0.389	87.0	2.8 ± 0.3	543.	2.87
T-126	1/4 × 1/4 in.	165	0.527	38.8	3.6 ± 0.3	49.0	0.79
T-606		—	0.092	0.71	79 ± 28		—
G-41		—	0.447	21.9	4.4 ± 0.1		—
G-52		—	0.436	27.4	3.9 + 0.2		—
G-56	1/2 × 1/2 in.	42	0.304	8.1	11.1 ± 1.1	84.	3.74
BASF	5 × 5 mm	87.3	0.500	11.8	7.3 + 0.7	41.	2.05
Harshaw	1/4 × 1/4 in.	44	0.489	13.3	7.2 ± 0.1	91.	3.95
Haldor Topsøe	1/4 × 1/4 in.	143	0.433	15.8[e]	2.8	25.8	0.83

Desig-nation	Percentage of Pore Volume in the Ranges (radius, Å)					Percentage of the Total Predicted Flux at Ambient Conditions from Pores in the Ranges (radius, Å)				
	<100	100–1000	1000–5000	5000–10 000	>10 000	<100	100–1000	10 00–5000	5000–10 000	>10 000
T-126	68	12	14	4	2	14	15	46	16	9
T-1258	76	5	10	5	4	17	7	36	20	20
T-826	65	9	5	8	13	10	6	13	26	45
T-314	68	10	14	6	2	12	12	44	23	9
T-310	76	8	11	4	1	20	13	42	21	4
G-39	75	6	6	10	3	10	8	22	46	14
G-35	75	6	6	10	3	10	8	22	46	14
T-606	0	62	26	5	7	0	36	38	11	15
G-58	0	61	26	10	3	0	33	41	20	6
T-126	68	12	14	4	2	9	15	42	23	11
T-606	27	40	20	4	9	5	25	36	10	24
G-41	74	10	12	3	1	11	14	50	18	7
G-52	45	47	5	1	2	12	56	18	2	12
G-56	45	45	7	1	2	7	54	22	6	11
BASF	61	29	10	0.1	0	10	45	44	1	0
Harshaw	56	37	7	0.25	0	18	49	31	2	0
Haldor Topsøe[g]	69	29	2	0	0	19	68	13	0	0

[a] Satterfield and Cadle [293].
[b] Diffusivity of hydrogen through nitrogen, measured at room temperature and atmospheric pressure, average of 5 sets of samples.
[c] τ_p calculated from Equation 1.48.
[d] τ_m calculated from Equation 1.33.
[e] Diffusivity of helium through nitrogen, multiplied by $\sqrt{4/2}$, average of 2 sets of samples.
[f] Prediction by parallel-path pore model, Equation 1.48.
[g] The predicted flux distribution for the Haldor–Topsøe catalyst in the original publication has a typographical error.

flux is obtained by integrating Equation 1.40 over the range of pore sizes, replacing $D_{12,\text{eff}}$ with $D_{12}\theta/\tau_p$:

$$N_1 = \frac{D_{12}\theta P}{\tau_p RT x_0(1 + N_2/N_1)} \times \int_0^\infty \ln\left[\frac{1 - (1 + N_2/N_1)(Y_1)_2 + D_{12}/D_{K1}}{1 - (1 + N_2/N_1)(Y_1)_1 + D_{12}/D_{K1}}\right] f(r)\,dr,$$

$$= \frac{-D_{\text{eff}} P}{RT x_0}(Y_2 - Y_1). \tag{1.48}$$

Here $f(r)\,dr$ is the fraction of the void fraction θ in pores with radii between r and $r + dr$. Thus, the flux is predicted by applying the relevant equation of mass transport to each pore size and summing the individual contributions over the whole pore-size distribution. The parallel-path pore model represented by Equation 1.48 has the important and useful characteristic of reducing to the simple Knudsen or bulk diffusion equations in its limits.

If a steady-state diffusion measurement is being made with two different gases on opposite sides of a pellet, but at identical total pressure, then the flux ratio N_2/N_1 is proportional to $\sqrt{M_1/M_2}$. The integration of Equation 1.48 may be readily carried out numerically by dividing the pore-size distribution into increments and calculating the average value of D_{12}/D_{K1} for each increment. The value of the bracketed expression in Equation 1.48 will drop with increasing values of D_{12}/D_{K1}, as one proceeds to smaller and smaller pores. In the Knudsen and transition range, only the larger pores make a significant contribution to the flux. The value of τ_p, which must exceed unity, may then be determined from Equation 1.48.

Johnson and Stewart [161] studied the counterdiffusion of nitrogen and hydrogen in a variety of samples of essentially pure alumina prepared in different ways to vary the pore-size distribution. The samples were formed into tablets and calcined at 900°F to convert the alumina to the γ form before measurements were made. The values of τ_p as calculated by this model varied between the extremes of 2.2 and 6.8.

Cadle [293] measured the effective diffusivities of seventeen different pelleted commercial catalysts and catalyst supports by the steady-state counterdiffusion method and analyzed his results (Table 1.13) by the parallel-path pore model described above. The catalysts are described in Table 1.14. The tortuosity factor calculated by this model, τ_p, was about four for half of the catalysts, and except for two materials that

Table 1.14 Catalysts and Supports in Table 1.13

Catalyst	Description
T-126	Activated γ-alumina
T-1258	Activated γ-alumina
T-826	3% CoO, 10% MoO_3, and 3% NiO on alumina
T-314	About 8–10% Ni and Cr in the form of oxides on an activated alumina
T-310	About 10–12% nickel as the oxide on an activated alumina
T-606	Specially compounded refractory oxide support
G-39	A cobalt-molybdenum catalyst, used for simultaneous hydrodesulfurization of sulfur compounds and hydrogenation of olefins
G-35	A cobalt-molybdenum catalyst supported on high-purity alumina, used for hydrodesulfurization of organic sulfur compounds
G-41	A chromia-alumina catalyst, used for hydrodealkylation and dehydrogenation reactions
G-58	Palladium-on-alumina catalyst, for selective hydrogenation of acetylene in ethylene
G-52	Approximately 33 wt % nickel on a refractory oxide support, prereduced. Used for oxygen removal from hydrogen and inert gas streams
G-56	A nickel-base catalyst used for steam reforming of hydrocarbons
BASF	A methanol synthesis catalyst, prereduced
Harshaw	A methanol synthesis catalyst, prereduced
Haldor Topsøe	A methanol synthesis catalyst, prereduced

had apparently been calcined at very high temperatures, the tortuosity factors all fell between 2.8 and 7.3. The macrovoid fractions for these commercial catalysts are substantially smaller than those reported in studies with laboratory pressed pellets. It is noteworthy that the same range of values of τ was found for bulk diffusion in commercial catalysts (Figure 1.5). In a similar study, Brown, Haynes, and Manogue [48] measured steady-state diffusion fluxes at atmospheric and higher pressures in twelve commercial pelleted catalysts and catalyst supports, including two of the same types studied by Cadle (Girdler T-126 and G-35). Except for one catalyst that had apparently been calcined at a very high temperature, the tortuosity factors as calculated by the parallel-path model all fell between 2.0 and 5.7, with a median value of three. The median value of τ_p, the extreme limits reported and the values for T-126 and G-35 are all about 25 per cent less than those reported by Cadle. The pore-size distributions as measured by mercury porosimetry on the same catalyst sample by different laboratories may vary by as much as 30 per cent [54] because of different values used for the surface tension and contact angle of the mercury, thereby affecting the calculated value of τ. In some cases the mercury porosimeter can only accommodate a fraction of a gram of catalyst, and it may be necessary to crush the pellets in order to obtain sufficiently small sizes to introduce into the porosimeter. This raises the possibility of sampling errors if the catalyst is not uniform in structure. However, in this case the differences seem to be primarily in the measurements of the fluxes themselves. The pore structure of a commercial catalyst may vary with pellet size even when the various sizes bear the same code designation.

For small molecules under room conditions, the transition region extends from about 100 to 10 000 Å. The percentage of the total flux (hydrogen and nitrogen) predicted by the parallel-path model to come from various pore ranges at the experimental conditions of Cadle's study are also given in Table 1.13. The flux is largely in the transition region, the contribution from the micropore region ($r < 100$ Å) does not exceed 20 per cent of the total in the most extreme case, although this region contributes the major portion of the pore volume in high area catalysts. As pressure is increased, however, the micropore region contributes increasingly to the total predicted flux. For example, at 65 atm, this region is calculated to contribute 50, 47, and 57 per cent of the flux for the Harshaw, BASF, and Haldor-Topsøe catalysts, respectively, as opposed to 19, 10, and 19 per cent of the flux at 1 atm. This emphasizes the importance of knowing the micropore-size distribution accurately when it is necessary to predict fluxes at high pressure.

In another investigation, Cadle [292] studied diffusion in five of these catalysts (three methanol synthesis catalysts, and G-58 and G-52) over the pressure range of 1 to 65 atm. The tortuosity factors in the parallel-path model were essentially invariant with pressure even though the dominant diffusion mode shifts from the lower end of the transition region up to essentially bulk diffusion as pressure is increased. The same result was found by Brown *et al.* [48] in their study of twelve commercial catalysts over a pressure range of one to 20 atm. Figure 1.11 plots the observed diffusion fluxes of helium through five commercial catalysts as a function of pressure [292]. The shapes of the curves are as anticipated from theory; the flux increases rapidly with an increase in pressure at low pressures and finally approaches a constant value at the highest pressure where diffusion is almost completely by the bulk mechanism.

Combining these results and the discussion in Section 1.6 on bulk diffusion in commercial catalysts, it is tentatively concluded that unless one is dealing with an unusual pore structure or is concerned with extreme experimental conditions, the effective diffusivity of a commercial porous pelleted catalyst or catalyst support can be predicted within a factor of about 2 using the parallel-path pore model (Equation 1.48) and a tortuosity factor of about 4. This degree of accuracy is probably

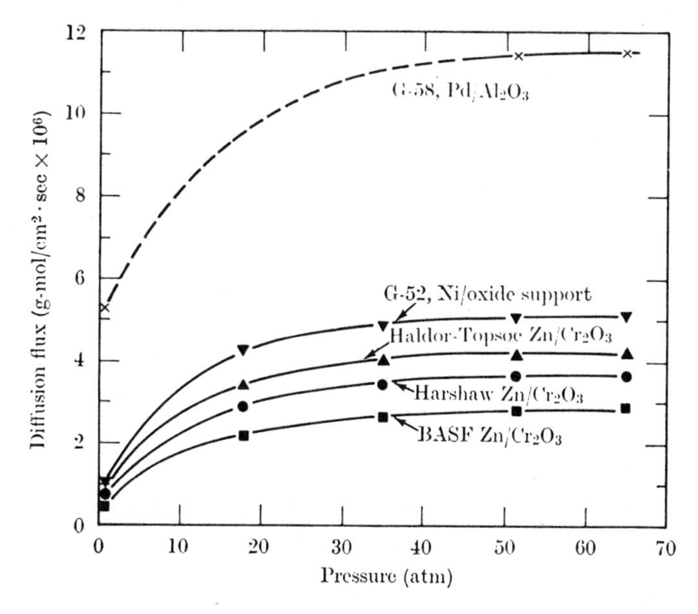

Figure 1.11. Helium flux through nitrogen as a function of pressure level. Satterfield and Cadle [292].

sufficient for most engineering applications. The use of more elaborate models may not improve the accuracy and merely increases the difficulty of calculation. The above generalization applies to reasonably uniform porous structures and to conditions in which surface diffusion or molecular-sieve-type diffusion is insignificant. The results indicate that a tortuosity factor obtained from a diffusion measurement under one set of conditions can be applied satisfactorily over a wide range of conditions even if the diffusion mode changes greatly. This conclusion is of great importance, since the temperature and pressure under reaction conditions are usually substantially different from those under which diffusion measurements are made.

In a few cases a significant pressure difference between the center and outside of a catalyst pellet may develop as a result of the change in number of moles on reaction (Chapter 4), whereupon mass transfer may occur by a process of simultaneous diffusion and forced flow. Relatively few experimental studies have been reported on this phenomenon under the conditions of interest in catalysis, namely finely porous solids and relatively small pressure difference. Otani, Wakao, and Smith [242] studied the counterdiffusion of hydrogen and nitrogen with a superimposed nitrogen pressure gradient through pressed boehmite (alumina) pellets and through porous Vycor. Evans, Truitt, and Watson [100] reported studies with helium and argon on a relatively large-pore carbon and Hewitt and Morgan [142] studies with oxygen and nitrogen on pile graphite. Cadle [292] reported studies with helium and nitrogen on a commercial methanol synthesis catalyst. Analysis and interpretation of results is complex and will not be pursued here. In terms of a parallel-path capillary tube model the tortuosity factor for the diffusion term is generally greater than for the flow term, the ratio of the two being about 1.8 for a methanol synthesis catalyst, 2 for the dense boehmite pellets and about 1.3 for the pile graphite [292].

1.8.2 *Anisotropic Structures*

In almost all cases diffusion measurements in porous catalysts or pressed pellets have been analyzed assuming the structure to be uniform, i.e., isotropic. However, two recent studies by Saraf [300] and Cadle [55] on pellets formed by pressing powder in an unlubricated die showed that the local effective diffusivity, as defined by Equation 1.42, varied with axial distance through the pellet by a factor of as much as $2\frac{1}{2}$ with 20 per cent chromia-on-γ-alumina, and 3 to 4 with boehmite alumina. The effect is caused by variation in the pore-size distribution with position in the pellet, which in turn arises from a maldistribution of pressure during the pressing, caused primarily by die-wall friction and to a lesser extent by interparticle friction. The effect would be expected to be magnified when diffusion measurements are in the Knudsen or transition range, as they were above, rather than in the bulk mode. Many measurements have been reported in the literature on compacted pellets of this type, and since they represent only average values through particular pellets they cannot be used with confidence to distinguish between alternate theoretical models of porous structures. Only scattered information is available on anisotropy in commercial catalysts and supports. Laboratory pressed pellets have frequently been of relatively large size (e.g., 1 in. diameter) and have been formed by pressing powders from one side only in an unlubricated die. Commercial pellets are generally much smaller, are usually pressed from both sides, and the usual incorporation of binders and lubricants in the catalyst mix should reduce die-wall friction. The gases evolved during the final calcination treatment should also help to modify any skin effect, so one may judge that commercially-pressed pellets should be much less anisotropic than laboratory-pressed pellets. This was borne out by studies on Girdler G-56, a nickel steam-reforming catalyst [54] in which the local diffusivity was essentially uniform with position and no skin effect was observed. Commercial spherical catalyst supports may be made by an agglomeration technique and appear in at least two cases [91, 298] to have had one or more fine-pore internal skins which in one instance caused the effective diffusivity for a gas to be perhaps about one-half of that otherwise obtainable.

The process of extrusion causes an orientation of particles and consequent anisotropic structures. Thus, the effective diffusivity of porous carbons formed by extrusion may be substantially different in the axial and radial directions (see discussion in Reference 54) and the same effect has been observed with kaolin-based catalysts. The outer layers typically have lower permeability—the so-called "skin' effect." Extruded catalysts, however, are seldom well formed and commonly there are numerous fine cracks in the curved surface from the calcination treatment which provide passageways into the interior. The degree of uniformity of catalysts and supports prepared by a dissolution process (e.g., porous Vycor glass, Raney nickel) or selective gasification — e.g., activated carbons, depends in part upon the extent to which concentration gradients (diffusion limitations) occurred during their preparation.

1.8.3 *Other Models of Porous Structures*

Wakao and Smith [355] proposed a "random pore" model that divides the pores into micropores and macropores and represents the diffusion flux as being the sum of that through the macropores, that through the micropores, and that by a series diffusion through both. The resulting diffusion equations represent elaborations of Equation 1.34 for application to bimodal pore-size distributions (Section 1.5.4).

The effective diffusivity in this model is given by

$$D_{\mathrm{eff}} = \frac{\theta_{\mathrm{macro}}^2}{\dfrac{1-(1+N_2/N_1)Y_1}{D_{12}} + \dfrac{1}{\bar{D}_{K,\,\mathrm{macro}}}} + \left[\frac{\theta_{\mathrm{micro}}^2(1+3\theta_{\mathrm{macro}})}{(1-\theta_{\mathrm{macro}})}\right] \times \left[\frac{1}{\dfrac{1-(1+N_2/N_1)Y_1}{D_{12}} + \dfrac{1}{\bar{D}_{K,\,\mathrm{micro}}}}\right] \tag{1.49}$$

The first term represents the flux through the macropores alone and the second that through the macropores and the micropores in series plus micropores alone. The separation of the pores into a distribution of macropores and one of micropores is somewhat arbitrary; generally about 100 or 125 Å radius has been taken as the upper limit of the micropore distribution. For evaluation of $\bar{D}_{K,\mathrm{macro}}$ and $\bar{D}_{K,\mathrm{micro}}$ an average pore radius is defined as

$$\bar{r} = \frac{\int_{v_1}^{v_2} r\, dv}{v_2 - v_1}. \tag{1.50}$$

For the micropores v_1 is 0 and $v_2 - v_1$ is the volume contained in pores up to 100 or 125 Å radius. For macropores, $v_2 - v_1$ represents the volume of pores above 100 or 125 Å radius. The use of $\bar{r}$ as defined by Equation 1.50 to calculate a diffusion coefficient is valid only if diffusion is completely by thè Knudsen mode over the entire pore size range, in which case D_K can be calculated from Equation 1.31. Diffusion in the macropores will usually be in the transition region, in which case there is no simple relationship between an average pore radius defined by Equation 1.50 and the effective diffusion coefficient; the latter becomes a function not only of the pore-size distribution, but also of pressure, temperature, and nature of the diffusing species. Thus, the use of $\bar{D}_{K,\mathrm{macro}}$ for extrapolation of data is not valid for most cases, but its contribution to D_{eff} varies greatly depending upon the pore structure and the conditions under which diffusion is occurring. For catalysts of relatively high porosity the first term in Equation 1.49 will be predominant for gaseous diffusion of small molecules at ambient conditions and higher pressures. For equimolal counterdiffusion Equation 1.49 then reduces to

$$D_{\mathrm{eff}} = \theta_{\mathrm{macro}}^2 \quad D_{12}. \tag{1.51}$$

This is equivalent to a tortuosity $\tau = 1/\theta_{\text{macro}}$, in the parallel-path pore model. Values of θ_{macro} generally fall between 0.2 and 0.6 (Figure 1.5), which corresponds to values of τ of 5 to 1.8. For solids possessing only a micropore structure, such as silica-alumina gel catalysts and porous Vycor glass, $\theta_{\text{macro}} = 0$, only the second term in Equation 1.49 will be significant, and diffusion will be usually by the Knudsen mode only. Equation 1.49 thus reduces to

$$D_{\text{eff}} = \theta_{\text{micro}}^2 \cdot D_{K1} . \tag{1.52}$$

Values of θ_{micro} are typically 0.3 to 0.5 for such systems (Tables 1.9 and 1.10) whence τ, which now equals $1/\theta_{\text{micro}}$, has values of 3 to 2. The Wakao-Smith model agreed well, without an adjustable parameter, with measurements at pressures from 1 to 12 atm on 1-in.-diameter $\times \frac{1}{2}$-in.-thick pellets pressed from boehmite alumina powder of various densities, but was somewhat less accurate for prediction for pellets of other sizes of pressed boehmite [53, 271, 274] and poor for porous Vycor [260]. In analysis of diffusion at atmospheric pressure in various commercial pelleted catalysts [48, 293], the degree of deviation between predicted diffusivities from the random-pore model and those observed was approximately comparable to the degree of variation of τ_p in the parallel-path model. However, the random-pore model was less reliable in predicting the effect of pressure level on diffusion flux than the simple assumption of a constant tortuosity factor in the parallel-path model. Most laboratory-pressed catalyst pellets and commercial pellets can be characterized by a tortuosity factor between extremes of about 2 and 7 in the parallel-path model, regardless of whether diffusion is in the bulk, Knudsen, or transition region. It appears that for the same structures, the random-pore model will likewise predict the effective diffusivity for small molecules at ambient temperature and pressure within a factor of about 2 to 3. For three commercial catalysts having a τ_p of 7, and porous Vycor with a τ_p of 6, the diffusivity predicted by the random-pore model was about twice that actually found and the deviations would be expected to be greater for those systems characterized by higher values of τ [293].

Butt and co-workers have recently proposed a relatively complicated computational model [107] and its application to a catalytic reaction [328]. The micropores are treated as dead space and the macropore structure as a series of diverging and converging arrays, each half being the mirror image of the other. The array comprises a series of pores of various lengths and diameters determined from the pore-size distribution and an adjustable parameter is introduced by the degree of mixing that is assumed to occur between the two types of arrays. In subsequent experimental studies [109] the counterdiffusion fluxes of helium and argon were measured at 1 to 15 atm and 0–69°C through laboratory-pressed pellets of a nickel oxide-on-kieselguhr catalyst and a molybdena catalyst. The results were interpreted in terms of the convergent-divergent model, but a general appraisal of its degree of validity has not as yet been made. Predictions by the Wakao-Smith random-pore model were about 30 per cent low for the nickel oxide catalyst but almost an order of magnitude low for the molybdena catalyst. Unfortunately, no comparisons seem to be available between the divergent-convergent model and the parallel-path model. Of the three models, the parallel-path model appears to be the most generally useful. It can be readily visualized physically, it seems to be reliable for extrapolation of data to other pressures, it is relatively easy to use, and in the absence of any diffusion measurements its predictive power is at least as good as that of any other presently available model.

For computational purposes some investigators have proposed a macro-micro structure in which the large pores are viewed as providing the passageways through the pellet and the micropores are visualized as branching off the macropores. This concept will be discussed in more detail in Section 4.6.2.

1.9 Additional Comments

At room temperature and atmospheric pressure, the ratio of diffusion coefficients in gases to those in liquids is of the order of 10 000. Thus a superficial analysis might suggest that with a liquid in the feed to a catalytic reactor the rate would be inordinately slower than for a gaseous feed, since the liquid would be absorbed by the porous catalyst, presumably making the internal surface ineffective. Reactor operating conditions, however, are such as to bring the two diffusion coefficients somewhat closer together. A typical value of $D_{K,\mathrm{eff}}$ at room temperature is 0.004 cm^2/sec and this increases as $T^{1/2}$. Taking D_{eff} for a liquid system to be 2.5×10^{-6} cm^2/sec at 25°C ($D_{12} = 10^{-5}$, $\theta = 0.5$, $\tau = 2$), and assuming activation energies of 0.77 and 4.3 kcal, the ratio of effective diffusion coefficients (gas/liquid) is found to drop from 1600 at 298°K to about 100 at 600°K (620°F).

Moreover, diffusion rates depend not only on diffusion coefficients but on gradients of concentration per unit volume. Solute concentration in a liquid is usually 50 to 500 times as great as for the same mixture as a vapor, the ratio depending on temperature and pressure. Thus, under reactor conditions, diffusion fluxes in pores filled with liquid can evidently be of the same order of magnitude as when only a gas phase is present.

The pressure drop accompanying gas flow in a fixed bed creates a pressure differential across each catalyst pellet. This causes a tendency for gas flow through the pellet. With small pores, rates of penetration of reactant by both flow and diffusion are proportional to the product of $D_{K,\text{eff}}$ and the ambient reactant concentration c_s. It is easily calculated that the flow due to the pressure drop is quite negligible compared to the diffusion of reactant in this situation. Displacement of fluid within the porous pellet may conceivably be significant for very large pores either with gas flow or with liquids flowing over the pellets.

2 Mass Transfer to Catalyst Particles

2.1 Introduction

For surface catalysis to be effective, the reactants present in the ambient fluid must be transported to the surface of the solid, and reaction products then transported back from surface to fluid. The diffusion path may be divided into two parts: bulk fluid to outer surface of the particle or pellet, and particle surface to active internal surface of the porous solid. This chapter deals with the resistance to transport between fluid and outer pellet surface. If the pellet is completely non-porous, this is the only diffusional resistance involved.

Industrial reactors may be conveniently classified, particularly for design purposes, as (1) those in which the solid catalyst particles remain in a fixed position relative to one another, and (2) those in which the particles are suspended in a fluid and are in motion.

The fixed bed is the common example of the first. A fixed bed over which liquid flows without filling the void spaces between particles is sometimes called a trickle bed. The fluidized bed, in which the catalyst particles are suspended in a gas stream, is the most common example of the second class. A slurry reactor is one in which the catalyst particles are suspended in a liquid. The fluidized bed is agitated by the flowing gas; slurry reactors are agitated by the flow of liquid, by the action of gas bubbling through the liquid, or by a mechanical agitator. Certain reactors are intermediate between the two classes noted above. Moving beds of pellets are sometimes used, as in some catalytic cracking processes. Though there is some motion of the pellets relative to each other, the reactor performance approximates that of a fixed bed.

Catalysts used in fixed beds are usually in the form of irregular granules, short cylindrical extrudates, or uniform spherical or cylindrical pellets. Mean particle diameters range from 1.6–6.4 mm ($\frac{1}{16}$–$\frac{1}{4}$ in.) in most fixed-bed applications. Fluidized beds employ much smaller particles, typically averaging about 50–75 μ in diameter for gas-solid systems.

Limited quantitative information is available regarding mass transfer from gases and liquids to particles in fixed and fluid beds. Representative data for these systems with single-phase fluid flow will be presented below.

2.2 Fixed Beds

2.2.1 *Mass Transfer*

Fluid passing over the surface of a pellet develops a boundary layer in which the velocity parallel to the surface varies rapidly over a very short distance normal to the flow. The fluid velocity is zero at the solid surface but approaches the bulk-stream velocity at a plane not far (usually less than a millimeter) from the surface. Mixing occurs in the bulk stream, and reactants and products are transported at rates that depend primarily on the nature of the flow.

Very near the surface, the fluid velocity is low and there is little mixing: Transport normal to the surface is by molecular diffusion. In the main fluid stream, mass transfer is essentially independent of the molecular diffusion coefficient D_{1m}, but very near the surface the rate is proportional to D_{1m}. As might be anticipated, therefore, the over-all process of transport between pellet and fluid is found experimentally to be proportional to D_{1m}^n, where n is between zero and unity. The flow passages in fixed beds are complex, and it has been found necessary to develop semiempirical correlations of data on mass transfer between the pellets and the flowing gas or liquid.

Data are commonly expressed in terms of a mass transfer coefficient k_c defined by

$$N = k_c(c_0 - c_s), \tag{2.1}$$

where N is the diffusion flux of the constituent in question, fluid to solid (g-mol/sec · cm²), c_s is the concentration at the surface, and c_0 is the concentration in the ambient fluid. In gas systems, the potential is taken to be the partial pressure of the diffusing substance and it is convenient to define a coefficient k_G by

$$N = k_G(p_0 - p_s), \tag{2.2}$$

where $k_G = k_c/RT$.

Dimensional considerations suggest the following as a basis for correlation of data on mass transfer:

$$N_{\text{Sh}} = \frac{k_c\, d_p}{D_{1m}} = f(N_{\text{Re}}, N_{\text{Sc}}), \tag{2.3}$$

where d_p is a representative dimension of the flow passage (taken to be the diameter or "equivalent diameter" of the pellets in the case of a fixed bed). The quantity on the left is termed the Sherwood number. The dimensionless Reynolds and Schmidt numbers are defined by

$$N_{\text{Re}} \equiv \frac{d_p G}{\mu}, \tag{2.4}$$

$$N_{\text{Sc}} \equiv \frac{\mu}{\rho D_{1m}}. \tag{2.5}$$

Here μ and ρ are the viscosity and density of the fluid, D_{1m} is the molecular diffusion coefficient for the diffusing species in the fluid, and G is the mass velocity of the fluid in (g/sec · cm ^{2}of total or superficial bed cross section normal to mean flow). The Schmidt number for gas mixtures seldom falls outside the range of 0.5–3; in liquids it is always greater than unity and varies over a range of several thousand fold in different systems. It has been shown experimentally that k_c is approximately proportional to $D_{1m}^{2/3}$ over a wide range of values of N_{Sc}. This indicates that, in Equation 2.3, $k_c d_p/D_{1m}$ should be proportional to $N_{\text{Sc}}^{1/3}$.

The length d_p appearing in N_{Re} is taken to be the particle diameter in the case of spheres; for cylinders d_p represents the diameter of a sphere having the same surface. If the length and diameter of the cylinder are x_c and d_c, then the value of d_p employed in N_{Re} is

$$d_p = [d_c x_c + \tfrac{1}{2} d_c^2]^{1/2}. \tag{2.6}$$

Knowledge of the total pellet surface in a fixed bed is necessary if bed performance is to be estimated from values of k_G or k_c. Table 2.1 lists values of pellet surface per unit volume of bed for the spheres and cylinders in the usual range of catalyst pellet sizes [319].

Table 2.1 Surface of Cylinders and Spheres in Fixed Beds [319]

d_p		Per Cent Void Space (100ε) 30	40	50
(cm)	(in.)	Values of a (cm^2/cm^3)[a]		
1.27	0.50	3.31	2.8	2.36
1.016	0.40	4.12	3.53	2.95
0.763	0.30	5.50	4.71	3.93
0.508	0.20	8.25	7.07	5.9
0.254	0.10	16.5	14.2	11.8
0.127	0.05	33.0	28.3	23.6

[a] Multiply area values (a) by 30.5 to obtain ft^2/ft^3.

Chilton and Colburn in 1934, as a basis for the correlation of mass transfer data, suggested the following relationship:

$$\frac{k_c \rho}{G} N_{Sc}^{2/3} = f(N_{Re}). \tag{2.7}$$

Since $c = P/RT$ in ideal gas mixtures, the equivalent expression involving k_G is

$$\frac{k_G P}{G_M} N_{Sc}^{2/3} = f(N_{Re}), \tag{2.8}$$

where P is the total pressure, and G_M is the molal velocity (moles mixture/sec $\cdot$ cm^2 of total bed cross section). These have been widely used to represent the results of experiments with many different systems, involving various geometries and flow conditions. The group on the left-hand side is usually symbolized by j_D; i.e.,

$$j_D \equiv \frac{k_c \rho}{G} N_{Sc}^{2/3} = \frac{k_G P}{G_M} N_{Sc}^{2/3}. \tag{2.9}$$

Studies of mass transfer from gases or liquids to pellets in fixed beds have been made by numerous investigators, most of whom reported their results as graphs of j_D or εj_D versus N_{Re}. The symbol ε represents the void spaces between pellets as a fraction of the total volume of the bed.

Data have been obtained by evaporating various liquids from porous pellets into gas streams, dissolving pellets of slightly soluble solids into flowing liquids, and extracting partially miscible liquids from porous pellets into flowing water. Correlations for gases at relatively low Reynolds numbers (e.g., below about 170) have been somewhat unreliable because mixing in an axial direction becomes increasingly significant (see later) and it is difficult to avoid essentially equilibrium conditions at the exit of even a short packed bed. Petrovic and Thodos [252] have recently presented new data obtained at low Reynolds numbers and have recalculated various earlier studies by Thodos and co-workers to correct the published data for axial mixing. Their recommended correlation, for $3 < N_{\mathrm{Re}} < 2000$, is

$$\varepsilon j_D = \frac{0.357}{N_{\mathrm{Re}}^{0.359}} \qquad \text{(gases)}. \tag{2.10}$$

For values of ε from 0.416–0.778 and a range of values of N_{Re} of 100–2000, Sen Gupta and Thodos [314, 315] found j_D to be inversely proportional to ε for gases in packed beds, hence the form of Equation 2.10.

Wilson and Geankoplis [391] have recently reported studies of mass transfer to *liquids* in packed beds and have reviewed earlier work. For $55 < N_{\mathrm{Re}} < 1500$ and ε between 0.35 and 0.75, they recommend

$$\varepsilon j_D = \frac{0.250}{N_{\mathrm{Re}}^{0.31}} \qquad \text{(liquids)}. \tag{2.11}$$

For $0.0016 < N_{\mathrm{Re}} < 55$, they recommend

$$\varepsilon j_D = \frac{1.09}{N_{\mathrm{Re}}^{2/3}} \qquad \text{(liquids)}. \tag{2.12}$$

Equation 2.12 was shown to correlate data for values of N_{Sc} over a range of 165–70 600. At the higher Reynolds numbers (e.g., exceeding about 55) the work of various investigators shows closely similar correlations for gases and liquids, the latter for a wide range of Schmidt numbers; i.e., the mass transfer coefficient is inversely proportional to $N_{\mathrm{Sc}}^{2/3}$ so that j_D is a function only of N_{Re}. The values of j_D as predicted by Equations 2.10 and 2.11 differ by only 15 per cent or less for $55 < N_{\mathrm{Re}} < 1500$. At low Reynolds numbers, however, j_D for liquids will be lower than j_D for gases. Mass transfer in packed beds is also discussed by Carberry [59], by Yeh [394], by Bradshaw and Bennett [44], and by the various investigators referred to in the referenced papers.

All of the foregoing relates to packed beds in which the flowing gas or liquid fills the voids between the particles: the fluid is a single phase.

Trickle beds, in which both gas and liquid occupy the voids, are discussed in Section 2.4.

2.2.2 *Heat Transfer*

Heat and mass are transferred between solid and fluid by similar mechanisms and data on heat transfer in fixed beds are correlated in the same way as data on mass transfer. Thus

$$j_H \equiv \frac{h}{c_p G} N_{\mathrm{Pr}}^{2/3}, \tag{2.13}$$

$$h \equiv \frac{q}{T_s - T_0}, \tag{2.14}$$

$$N_{\mathrm{Pr}} \equiv \frac{c_p \mu}{k}. \tag{2.15}$$

Here h is the heat transfer coefficient, q is the heat flux (per unit pellet surface area), c_p is the heat capacity per unit mass of fluid, N_{Pr} is the Prandtl number, T_s is the pellet surface temperature, T_0 is the fluid stream temperature and k is the thermal conductivity of the fluid.

For many geometries, j_H is approximately equal to j_D. DeAcetis and Thodos [93] measured heat and mass transfer rates by evaporation of water from beds of packed spherical particles, and reported the ratio of j_H/j_D to be 1.51, based on data for $15 < N_{\mathrm{Re}} < 2000$, but this was later revised to 1.076 [314] upon further evaluation of their data and analysis of that of others. Resnick [299] studied the heterogeneous decomposition of H_2O_2 vapor on passage through a bed of 6.4 mm solid silver spheres at partial pressures of H_2O_2 up to 110 Torr and total pressure of 1 atm (760 Torr) and Reynolds numbers up to 160. The reaction was completely controlled by mass transfer and temperature differences between reacting vapor and catalyst surface as high as 270°C were observed. Values of j_D and j_H could be determined simultaneously and independently directly from this reacting system. Values of j_D agreed closely with those found from nonreacting systems and the ratio of j_H/j_D was found to be 1.37.

It is often of considerable interest to be able to estimate the difference between the temperature of the catalyst pellet and that of the fluid in cases of highly exothermic or endothermic reactions. This requires a knowledge of the heat transfer coefficient, obtained by the methods outlined above.

2.3 Fixed-Bed Reactor Performance with Mass Transfer Controlling

Let z be the distance through a packed bed, ρ_B the bulk density of the bed, r the reaction rate in a differential section (g-mol reactant/sec · g catalyst), and Y the mole fraction of reactant in the fluid stream. Then for steady-state operation,

$$\rho_B r dz = -G_M dY. \tag{2.16}$$

If mass transfer from fluid to pellet is controlling, the concentration of reactant at the pellet surface will be negligible. In this limit, with the fluid a gas, Equation 2.2 gives

$$\rho_B r = k_G a(p - 0) = k_G a YP. \tag{2.17}$$

Combining the last two equations and integrating, we obtain

$$\ln \frac{Y_1}{Y_2} = \frac{k_G aP}{G_M} z. \tag{2.18}$$

Y_1 and Y_2 are the mole fractions of reactant in feed and product, respectively, for a bed height of z cm.

Replacing k_G by j_D from Equation 2.9 and using Equation 2.10 to relate j_D to N_{Re}, we obtain

$$\ln \frac{Y_1}{Y_2} = \frac{j_D az}{N_{\text{Sc}}^{2/3}} = \frac{0.357az}{\varepsilon N_{\text{Re}}^{0.359} N_{\text{Sc}}^{2/3}}. \tag{2.19}$$

This assumes that the physical properties of the fluid and the mass transfer coefficient do not change appreciably with bed length.

It is helpful to visualize the reaction length required for a specified degree of conversion in terms of a multiple of the particle diameter d_p. The product ad_p is independent of particle size and does not vary greatly over the fairly narrow range of values of ε encountered in the usual packed beds. Taking $\varepsilon = 0.4$ as typical, $ad_p = 3.6$ (Table 2.1) and Equation 2.18 becomes

$$\ln \frac{Y_1}{Y_2} = \frac{3.2z}{d_p N_{\text{Re}}^{0.359} N_{\text{Sc}}^{2/3}}. \tag{2.20}$$

N_{Sc} is nearly unity for gases. Equation 2.20 shows, for example, that for 99 per cent conversion ($Y_1/Y_2 = 100$) under mass-transfer-controlled conditions and with $N_{\text{Sc}} = 1$, the reactor length required is only about three particle diameters at $N_{\text{Re}} = 10$ and seventeen particle diameters at $N_{\text{Re}} = 1000$. For 90 per cent conversion and $N_{\text{Re}} = 1000$, z is eight particle diameters. If reactor lengths of approximately 50 to 100 particle diameters or more are required to go 90 per cent of the way to equilibrium with a first-order gas reaction, then the observed rate of reaction is far less than it would be under mass-transfer-controlled conditions and the concentration difference between gas stream and outer surface of catalyst pellets should be negligibly small. In liquid systems, N_{Sc} may be as large as 10 000 or more and Equation 2.20 shows that much greater bed depths are required than with gases to obtain the same conversion under mass-transfer-controlled conditions.

Mass-transfer-limited reactions occur industrially in the catalytic oxidation of ammonia-air mixtures on platinum gauze [95] as used in conventional nitric acid plants and in the closely related Andrussow process, in which a mixture of ammonia, air, and natural gas is converted to hydrogen cyanide on a similar catalyst. The catalytic decomposition of concentrated hydrogen peroxide on a silver catalyst [299] used for supplying auxiliary power in rockets, aircraft, and space vehicles, is likewise mass transfer controlled. Other examples include the combustion of carbon at high temperatures and the deposition of metals by decomposition of their compounds, as discussed in Section 1.1.

Example 2.1 Estimation of Coefficient of Mass Transfer from Gas Stream to Pellets in a Fixed Bed

A hydrodesulfurization reactor employs catalyst pellets in the form of cylinders 3.2 mm ($\frac{1}{8}$ in.) long and 3.2 mm ($\frac{1}{8}$ in.) in diameter. Operation is at 30 atm (425 psig) and at 660°K (728°F) with feed containing 82.8 mole per cent hydrogen and 17.2 mole per cent of 49.4°A.P.I. naphtha. The mixed vapor has a density of 0.0168 g/cm^3 (1.043 lb/ft^3), an average molecular weight of 30.3, and is fed to the reactor at a total rate corresponding to a mass velocity of 0.188 g/sec · cm^2 (1390 lb/hr · ft^2).

It is assumed that thiophene is representative of the small quantities of sulfur compounds to be removed. Using the data of Maxwell [208] and available procedures for estimating physical properties of mixtures [264], we estimate the viscosity of the gas mixture to be 0.00038 P, and N_{Sc} for thiophene in this gas mixture to be 2.96. From Equation 2.6, the equivalent sphere diameter d_p is 0.392 cm.

$$N_{Re} = \frac{d_p G}{\mu} = \frac{0.392 \times 0.188}{0.00038} = 194.$$

Estimating $\varepsilon \approx 0.4$, j_D is given by Equation 2.10,

$$j_D = \frac{0.357}{\varepsilon N_{Re}^{0.359}} = \frac{0.357}{(0.4)(6.6)} = 0.135,$$

$$j_D = 0.135 = \frac{k_G PM}{G} N_{Sc}^{2/3} = \frac{k_G \times 30 \times 30.3}{(0.188)} \times (2.96)^{2/3},$$

whence, $k_G = 1.35 \times 10^{-5}$ g-mol/sec·cm²·atm or 0.099 lb-moles/hr·ft²·atm.

The value of k_G found in this example must be translated into reactor performance in order that its significance be understood. Equation 2.18 becomes

$$\ln \frac{Y_1}{Y_2} = \frac{k_G aP}{G_M} z = \frac{1.35 \times 10^{-5} \times 30 \times 10.2 \times z}{(0.188/30.3)}.$$

Here a has been taken from Table 2.1 as 10.2 cm²/cm³, corresponding to a void volume of 40 per cent.

The following relation between sulfur removal and bed height is obtained by substitution in Equation 2.20.

Y_1/Y_2	2	10	100	1000
Percent sulfur removed	50	90	99	99.9
Bed depth (cm)	1.05	3.46	6.95	10.5

If sulfur removal depended only on mass transfer from gas to pellet, required bed heights would evidently be much less than those commonly used.

2.3.1 *Effects of Space Velocity*

Resistance to mass transfer between fluid stream and outer pellet surface is markedly reduced as the fluid velocity is increased. An observed increase in reaction rate with increase in fluid velocity at constant temperature indicates that mass transfer to the catalyst is limiting the reaction rate provided that a "differential reactor" is used; i.e., one in which the fractional change in reactant concentration through the system is minimal. Such a test can be carried out by use of a single catalyst pellet, a single wire, or a screen, as in the examples cited in Section 1.1. Increasing the velocity in a packed bed of catalyst is inconclusive in most cases. If the exit reactant concentration differs appreciably from the inlet, an increase in gas flow rate will increase the intrinsic rate of reaction by increasing the average concentration, entirely aside from any effects on mass transfer as such and it is difficult to disentangle the two effects unless the intrinsic kinetics are simple and known — conditions seldom encountered. This can be circumvented by studies on catalyst beds of two different depths. At the same contact time (space velocity), the linear velocity will be greater in the deeper bed; and, in the absence of mass transfer limitations, the concentration profile and degree of conversion will be the same for the two cases. By this method, for example, Dowden and Bridger [96] were able to show the absence of bulk mass transfer limitations in the oxidation of SO_2 in air to SO_3 on a commercial vanadium pentoxide catalyst at 400° and 470°C. Other precautions, however, must be taken. If bulk mass transfer is indeed significant, the analysis in Section 2.3 shows that a substantial degree of conversion will occur in a very short depth or even possibly on a single catalyst pellet. With a highly exothermic reaction interpretation of the results may be difficult since axial and radial temperature gradients may be severe and ignition and quenching phenomena may develop.

For flow through a packed bed, dispersion in the axial or radial direction is characterized by the Peclet number $N_{Pe} = d_p u / D_T$, where u is the average linear velocity in the particle interstices and D_T is a dispersion coefficient (cm^2/sec) analogous to a diffusion coefficient. It is obtained by assuming that dispersion follows Fick's law regardless of the actual mechanism of transport. The mechanism of dispersion is different in the axial and radial directions except for the limiting case at very low Reynolds numbers, where dispersion is by molecular diffusion only. Hence, for a particular value of the Reynolds number, the value of D_T and the corresponding Peclet number will differ in the two directions. The Peclet number for *radial* dispersion reaches an asymptotic value of about 12 for both gases and liquids at Reynolds numbers above about 100. The *axial* Peclet number reaches an asymptotic value of about 2 at a Reynolds number above about 10 for gases and above somewhat higher Reynolds numbers for liquids. The axial Peclet number of 2 signifies that each fluid region between catalyst particles behaves like a well-mixed stage. Each particle length is approximately equivalent to one completely stirred reactor, so a packed bed exceeding 20 to 30 pellets in depth behaves in this flow region essentially as a plug-flow-type reactor. Studies of reactions in the intermediate flow regime may be very difficult to interpret. Mass transfer rates are low and the nature of the flow can give rise to unknown concentration, temperature, and residence time distributions which can change substantially with slight variation in experimental conditions. The hydrodynamic regions that exist in packed-bed reactors are discussed in an excellent review by Wilhelm [389].

Scaleup of reactors is usually done in terms of "space velocity," variously defined as the volumetric flow rate under reactor conditions divided by the bed volume, the flow rate (mass or volume) per gram or pound of catalyst, or as the cold liquid volumetric feed rate divided by the bed volume. The last, abbreviated as LHSV (liquid hourly space velocity) is based on the liquid feed volume, even though this may be vaporized and mixed with other reactants before entering the catalyst bed. It is evident that "space velocity" has units of reciprocal time when the basis is bed volume.

An industrial reactor is usually substantially longer than a laboratory or pilot plant reactor. For a given space velocity the "mass velocity," mass flow rate per unit total cross-sectional area of the bed, will be greater in the industrial reactor so bulk mass transfer limitations will tend to be less important in the industrial unit than in the smaller unit. The space velocity is a useful concept for reactors approximating plug flow provided that the "holdup," or average residence time, does not vary appreciably with reaction conditions or between laboratory and plant reactors. This may not be the case when a gas and a liquid are being reacted. In a trickle-bed reactor (Section 2.4) the "holdup" and actual contact time for a given space velocity with respect to the liquid can be markedly affected by the gas flow rate.

For a given mass velocity, a change in reactor diameter also changes the radial temperature profile if heat is being transferre dto or from the walls. The maximum reactor diameter is frequently set by the maximum radial temperature gradient allowable to achieve the desired yield or conversion. Analysis of the interaction between intrinsic kinetics, temperature and concentration gradients, and reactor diameter is a complex problem in the design of fixed-bed reactors and is outside the scope of this book.

2.3.2 *Temperature Difference between Solid and Fluid*

Estimation of the temperature difference, fluid to pellet surface, is important in connection with the analysis of data on catalytic reactions involving substantial heat effects.

Example 2.2 Estimation of Temperature Difference between Stream and Pellet in a Fixed Bed

In the hydrodesulfurization case described in the preceding example, what is the temperature difference, pellet surface to gas, at a point in the bed where the heat release is 0.0247 g-cal/sec · cm³ or 10 000 Btu/hr · ft³ bed?

Assuming c_p to be 0.9 g-cal/g · °C and k to be 5.4×10^{-4} g-cal/sec · cm² · °C/cm, or 0.13 Btu/hr · ft² · °F/ft for the gas mixture, N_{Pr} is 0.63; N_{Re} is 194, as in the earlier example. Taking $j_H = j_D$ and substituting in Equations 2.13 and 2.14, we obtain

$$\begin{aligned} h &= 0.135 \times 0.9 \times 0.188 \times (0.63)^{-2/3} \\ &= 0.031 \text{ g-cal/sec} \cdot \text{cm}^2 \cdot {}^\circ\text{C} \quad \text{or} \\ &\quad 230 \text{ Btu/hr} \cdot \text{ft}^2 \cdot {}^\circ\text{F}, \end{aligned}$$

$$T_s - T_o = \frac{0.0247}{0.031 \times 10.2} = 0.078°\text{C or } 0.14°\text{F}.$$

This temperature difference is negligibly small. With a mass velocity of 0.188 and a fluid specific heat of 0.9, the assumed heat release corresponds to a rise in temperature of the gas stream of 0.0247/0.188 × 0.9 or 0.146°C per cm (8°F per foot of length) in the direction of flow in the bed, which is not uncommon in hydrogenations.

The relationship between the degree of mass transfer control of a reaction and temperature difference between pellet surface and fluid may be easily derived for steady-state conditions from Equations 2.1 and 2.14. The rate of mass transfer of a reacting species from fluid to solid multiplied by the heat of reaction per mole of diffusing species must equal the rate of heat transfer from solid back to fluid. Hence

$$k_c(c_0 - c_s)(-\Delta H) = h(T_s - T_0). \tag{2.21}$$

Substituting the expressions for the Prandtl and Schmidt numbers, Equations 2.15 and 2.5, and for the j_D and j_H functions, Equations 2.9 and 2.13,

$$(T_s - T_0) = \left(\frac{j_D}{j_H}\right)\left(\frac{N_{\text{Pr}}}{N_{\text{Sc}}}\right)^{2/3} \frac{(-\Delta H)}{\rho c_p}(c_0 - c_s). \tag{2.22}$$

The extent to which the reaction is bulk mass transfer controlled, f, may logically be defined as the ratio $(c_0 - c_s)/c_0$, whence Equation 2.22 becomes

$$(T_s - T_0) = \left(\frac{j_D}{j_H}\right)\left(\frac{N_{\text{Pr}}}{N_{\text{Sc}}}\right)^{2/3} \frac{(-\Delta H)c_0}{\rho c_p} f. \tag{2.23}$$

The temperature difference is seen to be directly proportional to the heat of reaction per mole of diffusing component and to the fractional drop in concentration between bulk fluid and solid. The product $(-\Delta H)c_0$ is the heat that would be released by complete reaction of 1 cm^3 of reactant mixture, and the product ρc_p is the volumetric heat capacity of the gas. The quotient $(-\Delta H)c_0/\rho c_p$ represents the temperature rise that would be calculated for complete adiabatic reaction of the fluid mixture.

For many simple gas mixtures, the ratio $N_{\text{Pr}}/N_{\text{Sc}}$ is in the vicinity of unity, as is also j_D/j_H. For a completely mass-transfer-controlled gas-phase reaction ($f = 1$) corresponding to the above circumstances, the temperature difference between gas phase and solid will thus be approximately equal to the calculated adiabatic temperature rise for complete reaction of the fluid. (This assumes heat transfer by radiation to be negligible, which may not be the case for some kinds of reactors, particularly at high temperatures.)

An example of such a reaction is the industrial oxidation of ammonia-air mixtures on a platinum gauze to form nitric oxide [231]. The gas mixture is at essentially ambient temperature when it enters the reactor, but the gauze reaches temperatures of 750–900°C. Equation 2.22 shows that it is possible for the catalyst surface temperature to exceed the adiabatic reaction temperature. Equation 2.23 emphasizes the fact that, if the heat of reaction is large, mass transfer limitations may be small, yet heat transfer can still be important. Consider, for example, a case in which the calculated adiabatic temperature rise for the reaction is 500°C and conditions at a point in a reactor correspond to only 4 per cent concentration difference between bulk gas and catalyst, i.e., $f = 0.04$. Taking $(j_D/j_H)(N_{Pr}/N_{Sc})^{2/3}$ as unity, the temperature difference would be 20°C, sufficient to cause a marked increase in the observed rate of reaction over that which occurs if the catalyst were indeed at the vapor temperature. With sufficiently active catalysts a region of unstable catalyst temperatures exists. For example, upon increasing the temperature of the reacting gas, the temperature of the catalyst will suddenly jump to a new, higher level. The situation is analogous to the ignition of a fuel such as carbon in a stream of air. Theoretical analyses of this effect have been published [110, 299, 351], and it has been demonstrated experimentally in the catalytic decomposition of hydrogen peroxide vapor [299] and the catalytic reaction of hydrogen and oxygen on platinum [90]. This is one category of instability phenomena that appear in a variety of situations involving simultaneous diffusion and chemical reaction. Others are discussed in Chapters 4 and 5.

With a rapid exothermic catalytic reaction of a liquid forming a gas or vapor as a product, phenomena may be observed similar to the transition from nucleate to film-type boiling in heat transfer, as shown in a study of the decomposition of aqueous solutions of hydrogen peroxide on silver metal in the form of cylinders and wires of various diameters [291]. The catalyst temperature exceeds the liquid temperature and the rate of reaction shows a maximum as H_2O_2 concentration is increased. At low concentrations oxygen bubbles are observed to evolve over the entire catalyst surface but above concentrations of 60 wt % or higher, gas surrounded the surface on all sides. At these high concentrations, the rate of reaction could be *increased* by a factor of as much as 7 by *cooling* the catalyst. As with ordinary film-type boiling the maximum rate of reaction obtainable increased substantially the smaller the diameter of the cylinder or wire catalyst.

The use of mass transfer and heat transfer correlations developed from nonreacting systems would be expected to apply to reaction between fluid and solid only if reaction occurs completely at the solid-fluid interface. Under some circumstances this may not occur. One would anticipate that in some hydrocarbon oxidations and similar reactions, it would be possible to have long-chain free-radical reactions occurring homogeneously, initiated by free radicals generated on the catalyst surface. In such cases the observed rate of reaction might be substantially greater than that calculated for complete mass transfer control, even if this were indeed not the case. Similar behavior may be expected in some liquid-phase systems.

In all the above discussion, the mass transfer coefficients to the solid particles are the averaged values around the interface. It is evident that the flow characteristics and, therefore, the point mass transfer coefficient from a fluid to a solid particle will vary with position around the particle. At low flow velocities fluid motion is laminar, and each element of the fluid follows a definite path that, in principle, is known or can be calculated. Mass transfer from a fluid in laminar flow to a solid surface can, again in principle, be calculated, though the analysis is often difficult. The equations are the simplest for mass transfer between the surface of a rotating disk whose axis is perpendicular to the plane of the disk, and a fluid in laminar motion. The problem was first analyzed by von Karman in 1921, and the analysis is given by Levich [189]. The relation is

$$\frac{2k_c R}{D_{1m}} = 1.24(N_{\mathrm{Sc}})^{1/3}\left(\frac{\rho\omega R}{\mu}\right)^{1/2}, \tag{2.24}$$

where D_{1m} is the molecular diffusion coefficient, R is the radius of the disk, and ω is the angular speed of rotation. The quantity on the left is the Sherwood number (also termed the Nusselt number for mass transfer), and the dimensionless ratio $\rho\omega R/\mu$ is a version of the Reynolds number. The three-dimensional flow pattern causes k_c to be the same at all points on the surface. The rotating disk and the very long rotating cylinder are examples of the relatively few cases in which the surface is "uniformly accessible," which greatly simplifies the analysis of experimental data, with or without surface catalysis. The rotating disk system has long been used in electrochemical studies but only recently has it apparently been adapted to studies between a solid and a gas [232].

Laminar flow over a flat plate is a case in which the surface is not uniformly accessible: The local value of k_c decreases sharply with distance

x from the leading edge. The average value of k_c, from which the mass transport to the entire plate may be calculated, is given by [189, 253]

$$\frac{k_c x}{D_{1m}} = 0.66(N_{Sc})^{1/3}\left(\frac{\rho x U}{\mu}\right)^{1/2}, \tag{2.25}$$

where x is the length of the plate in the direction of flow, and U is the ambient velocity of the fluid. Rosner [276] presents the results of calculations that show how mass transfer limitations in this system affect the observed kinetics. This geometry is of much interest in aeronautical engineering, but the marked variation in mass transfer coefficient with position makes it difficult to study cases in which the mass transfer is combined with chemical reaction.

More theoretical background together with examples of specific studies are given in the book by Frank-Kamenetskii [111] and in a review by Bircumshaw and Riddiford [30].

There have been several theoretical analyses of flow within a tubular reactor involving the mathematical solution of the general diffusion equations with the surface rate as the boundary condition [86, 153], and the theory has been applied to experimental studies of the catalytic oxidation of SO_2 [17], catalytic oxidation of ammonia [162], and the catalytic decomposition of hydrogen peroxide [301].

2.4 Trickle Beds

Fixed-bed catalytic reactors are sometimes employed in cases where the feed consists of both a gas and a liquid. The liquid is usually allowed to flow down over the bed of catalyst, while the gas flows either up or down through the void spaces between the wetted pellets. Cocurrent downflow of the gas is generally preferred because it causes much better distribution of liquid over the catalyst bed and higher liquid flow rates are possible without flooding.

Such trickle-bed reactors have been developed by the petroleum industry over the past ten to fifteen years for hydrodesulfurization, hydrocracking, and hydrotreating of various petroleum fractions of relatively high boiling point. Under reaction conditions, the hydrocarbon feed is frequently a vapor-liquid mixture which is reacted at liquid hourly space velocities (LHSV in volume of fresh feed, as liquid/volume of bed · hr) in the range of 0.5–4, in the case of hydrodesulfurization. Both direct costs and capital costs are claimed to be 15–20 per cent less for the trickle-bed operation than for the equivalent hydrodesulfurization unit operating entirely in the vapor phase [154]. Furthermore, it is desired to process many stocks of such high boiling point that vapor-phase operation would be impossible without excessive cracking, as for example, in the refining of lubricating oils by treatment with hydrogen or hydrodesulfurization of residual stocks. One can foresee the extension of application of trickle-bed reactors to other gas-liquid reactions as more becomes known about their design and operating characteristics. Unlike the slurry reactor, a trickle-bed reactor approaches plug-flow behavior and the problem of separating the catalyst from the product stream does not exist. The low ratio of liquid to catalyst in the reactor minimizes the extent of homogeneous reaction. On the other hand, if reaction is substantially exothermic, the heat evolved may cause portions of the bed not to be wetted with consequent poor contacting of catalyst and liquid.

The use of liquid phase in the feed introduces several problems. The liquid distribution over the catalyst varies greatly with liquid and vapor flow rates, the properties of the reactants, the design of the reactor and of the liquid distributor. These affect the contacting efficiency between liquid and catalyst and the liquid holdup in the bed. Gaseous reactants must first be absorbed and transported across a liquid film to the outside catalyst surface and then through liquid-filled pores to the reactive surface.

Very little information is available regarding mass transfer limitations in trickle beds. In some laboratory studies of trickle-bed reactors the rate of reaction appeared to be limited by intrinsic reaction kinetics rather than a mass transfer process. LeNobel and Choufoer in a study of the performance of a laboratory trickle-bed reactor representative of the Shell hydrodesulfurization process [188], found that reduction in catalyst size increased the reaction rate, indicating pore-diffusion limitations. Reduction in the feed liquid viscosity by addition of a diluent also increased the reaction rate; this was attributed to an increase in diffusivity of organosulfur molecules through the liquid-filled pores. The Shell process was discussed further by van Deemter [343], who concluded that under typical industrial reaction conditions (55 atm pressure, 370°C) the effectiveness factor (see Section 3.1) was about 0.36 on $Co/Mo/Al_2O_3$ catalyst particles about 5 mm in diameter. Adlington and Thompson [2], from studies at the British Petroleum Company, concluded that at 416°C and 34 atm pressure, the effectiveness factor was about 0.6 for $\frac{1}{8}$-in. pelleted catalyst of the same type. The liquid film around catalyst pellets has an average thickness of 0.01–0.1 mm under typical hydrodesulfurization conditions, which is so much less than the radius of the usual catalyst particles that it will not constitute a significant mass transfer resistance unless the reaction rate is so fast that the effectiveness factor for reaction *inside* the catalyst pellets is extremely low. Van Zoonen and Douwes [346] studied the hydrodesulfurization of a straight-run gas oil on 3 × 3 mm pellets of a 3.6 per cent CoO/10.4 per cent MoO_3/alumina catalyst at 34 atm pressure and 375°C, and on the same catalyst after crushing to form fine particles. For a fixed liquid flow rate in a reactor of fixed size the degree of desulfurization was greater on the crushed catalyst than on the pellets, and greater on pellets deliberately formed to have low density (and, therefore, high void fraction) rather than on higher-density pellets, even though the total surface area of catalyst in the reactor was substantially less when packed with the lower-density material. The effectiveness factor on 3 × 3 mm pellets was estimated to range from about 0.5 on high-density pellets to 0.8 on low-density pellets. Back-calculation of the effective diffusivity from reaction rate data gave values of about 10^{-8} cm^2/sec, which are so low that the pores must have been filled with liquid rather than vapor, even though more than 95 per cent of the feed was believed to be in the vapor phase.

In most cases of trickle-bed operation the catalyst pores are presumably filled with liquid unless a gas is formed as a product and the effectiveness factor can be estimated by the procedures discussed in Chapter 3. With a rapid exothermic reaction, pores may be filled with vapor at bulk liquid temperatures somewhat below the boiling point of products or reactants. An example seems to exist in the catalytic hydrogenation of benzene to cyclohexane, in which about 49.8 kcal are evolved per mole of cyclohexane formed, as reported by Ware [359] for studies in a trickle-bed reactor on a commercial Ni-on-kieselguhr catalyst in the range of 30–120°C and 7.2–34 atm pressure. Over a certain range of conditions (about 70–100°C) two separate reaction rates were found for any specific set of steady-state operating conditions, depending upon the method used to start up the reactor. One rate was 5 to 10 times greater than the other. The higher rate was interpreted as a system in which pores were filled with benzene vapor and hydrogen and mass transfer limitations were negligible; the lower rate was interpreted as involving diffusional resistance inside liquid-filled catalyst pores.

Pelossof [298] studied the hydrogenation of α-methylstyrene to cumene at 20–50°C and 1 atm on a trickle bed reactor comprising a single vertical column of spherical porous Pd/alumina catalyst pellets; he also reported information on liquid holdup in this system. This reaction at 50°C is much faster than hydrodesulfurization as carried out under industrial conditions, and the reaction rate was about one-half of that which would be expected in the absence of mass transfer limitations in the outside liquid film, which varied from 0.01–0.02 cm in thickness. In a trickle bed the concentration at the outside pellet surface and hence mass transfer rate through the film is affected by the internal effectiveness factor, which here for the pellets alone was 0.006.

Four different models for the film resistance are presented by Pelossof and discussed. At low flow rates a simple stagnant film model may be used for first-approximation calculations for the value of k_{Ls} for mass transfer across the film. At high flow rates a considerable amount of mixing of the liquid in the film is presumed to occur at each point that liquid transfers from one catalyst particle to the next, and this increases the value of k_{Ls}. It is noteworthy that even with substantial mass transfer resistance through the liquid film, the observed rate of reaction for fixed average reactant concentration in the bulk will change little with liquid flow rate, partly because the film thickness Δ is proportional to the cube root of flow rate, and also because better mixing occurs at the points of contact of catalyst particles at higher flow rates.

Following is a suggested procedure for determining if, in trickle-bed reactors, mass transfer through the flowing liquid film may be a significant resistance. Assuming that no mass transfer resistance exists within the gas phase, then by a mass balance on hydrogen (or other gas) diffusing through the liquid film around a single cylindrical pellet (diameter is d_p and length d_p):

$$\left(\frac{-1}{V_c}\frac{dn}{dt}\right)\frac{\pi d_p^3}{4} = k_{Ls}\left(\frac{\pi d_p^2}{2} + \pi d_p^2\right)(c^* - c_s), \tag{2.26}$$

where c^* is the saturation concentration of the gas in the liquid. Simplifying,

$$\frac{d_p}{6k_{Ls}c^*}\left(\frac{-1}{V_c}\frac{dn}{dt}\right) = 1 - \frac{c_s}{c^*}. \tag{2.27}$$

A criterion for significant mass transfer limitations through the liquid film can be said to correspond to $c_s < 0.95\ c^*$. Application of this criterion to Equation 2.27 leads one to the conclusion that mass transfer from the gas phase to the outer surface of the catalyst pellets will not be significant unless the inequality of Equation 2.28 holds;

$$\frac{10\ d_p}{3c^*}\left(\frac{-1}{V_c}\frac{dn}{dt}\right) > k_{Ls}. \tag{2.28}$$

A conservative estimate of k_{Ls} when no information is available on either k_{Ls} or $\bar{\Delta}$, the average film thickness, is obtained by taking it to be $2Da/\varepsilon$. The maximum mass transfer resistance is Δ/D. The average film thickness may be approximated as $\bar{\Delta} = \varepsilon/2a$ by assuming that 50 per cent of the voids in the trickle-bed reactor is filled with a liquid [277], where a is the outside area of catalyst particles per unit volume of reactor. For example, for $\frac{1}{8}$-in.-diameter pellets, this gives a film thickness of about $\frac{1}{3}$ mm.

The average film thickness $\bar{\Delta}$ can be estimated from correlations for dynamic hold up in packed beds such as that of Shulman and co-workers [321], although these must be used cautiously since the data were obtained primarily on packings, such as Raschig rings and Berl saddles, rather than typical catalyst particles. The published data were also obtained at the higher flow rate conditions characteristic of absorption towers rather than trickle-bed reactors.

A value of k_{Ls} may also be calculated [298] by applying a form of penetration theory which might be expected to be valid at short contact times and high flow rates. If the two concentration boundary layers do not overlap appreciably during the passage of liquid over one catalyst particle then $1/k_{Ls} = 1/k_L + 1/k_s$, where k_L is the absorption mass transfer coefficient and k_s is that for the dissolution of sparingly soluble particles in a liquid film. Correlations of experimental values are given, for example, by van Krevelen and Krekels [344] for k_s and by Sherwood and Holloway [318] for k_L. However, absorption mass transfer coefficients have usually been obtained for packings that are different in size and shape from those commonly used in catalytic reactors and under substantially different operating conditions, so values are somewhat uncertain when used for the present purpose.

Example 2.3 Application of Criteria for Absence of Mass Transfer Limitations through the Liquid Film of a Trickle-Bed Reactor

Consider a set of data on the Shell trickle-bed hydrodesulfurization process [188]. At 367°C, 827 psia pressure, and space velocity of 2.0 kg/liter of bed · hr, on $\frac{1}{8}$-in. pellets, ($d_p = 0.318$ cm), the reaction rate $(-1/V_c)(dn/dt)$ was 1.04×10^{-6} g-mol H_2/sec · cm³ pellet. The value of ε is 0.36; a is 9.87. We estimate the equilibrium hydrogen concentration (solubility) in the feed stock as 4.84×10^{-4} g-mol/cm³, the bulk diffusivity as 5.5×10^{-4} cm²/sec. The left-hand side of Equation 2.28 becomes 2.46×10^{-3} cm/sec. A value of k_{Ls} may be calculated applying the penetration theory outlined and correlations [318, 344] for k_L and k_s. This gives $k_{Ls} = 1.55 \times 10^{-2}$ cm/sec, and liquid film resistance is concluded to be negligible. This value is only approximate because the correlations for k_L and k_s are for different kinds of systems and it is uncertain that penetration theory can be applied at these low flow rates. However, k_{Ls} estimated as $2Da/\varepsilon$ is about 3×10^{-2} cm/sec, and the conclusion is the same.

As a second example, consider the data of Pelossof [298] for the hydrogenation of liquid α-methylstyrene to cumene on a single vertical column of spherical porous Pd/alumina catalyst pellets at 50°C and 1 atm hydrogen pressure. The observed rate of reaction was 1.86×10^{-7} g-mol/sec · cm³ of pellet, essentially independent of liquid flow rate; d_p was 0.825 cm; c^* was 3.54×10^{-6} g-mol H_2/cm³ liquid. The left-hand side of Equation 2.28 is 0.145 cm/sec. For the lowest liquid flow rate studied (0.133 cm³/sec, over a single column of spheres), $\bar{\Delta}$ was 0.012 cm. The bulk diffusivity of H_2 in α-methylstyrene is 1.65×10^{-4} cm²/sec, and if we take $k_{Ls} = D/\bar{\Delta}$, it equals 1.4×10^{-2} cm²/sec, which is much less than the left-hand side of Equation 2.28. The conclusion is that mass transfer limitations through the film are significant, and this was confirmed by experiment. In principle, averaging should be done with respect to $1/\Delta$ rather than Δ, although for laminar flow around a column of spheres the difference is not great. The value of $1/\overline{1/\Delta}$ here was 12 per cent greater than $\bar{\Delta}$. For this hydrogenation reaction at 50°C the intrinsic reaction rate constant was quite large, k_v being 16.8 cm³ fluid/cm³ catalyst pellet · sec in contrast to the pseudo-first-order rate constant of about 2.3×10^{-3} sec^{-1} typical of trickle-bed hydrodesulfurization.

With activities typical of Co/Mo/alumina catalyst under industrial hydrodesulfurization conditions, the diffusional resistance of the liquid film cannot be expected to be of importance. The study of van Zoonen and Douwes showed that even though diffusional resistances inside 3 x 3 mm catalyst pellets are significant with respect to hydrodesulfurization, nitrogen removal was not diffusion limited, nor was the rate of coke formation affected by particle size or porosity.

In some cases it appears desirable to contact a fixed bed of catalyst with liquid and gas in a cocurrent upflow pattern rather than downflow. The larger liquid holdup permits somewhat longer liquid contact time in a reactor of fixed size and may permit better contact between liquid and solid as well as better control of temperature if the reaction is substantially exothermic. Liquid distribution is improved by the increase in holdup. However, very little information seems to be available on the residence-time distributions to be expected in such a system, or the potential rates of mass transfer from gas phase to catalyst.

2.5 Fluidized Beds

Fluidized-bed catalytic reactors were first employed on a large scale for the cracking of petroleum stocks, but are now being used for an increasing variety of chemical operations, such as the partial oxidation of naphthalene or ortho-xylene to phthalic anhydride and the formation of acrylonitrile from propylene, ammonia, and air. Fluidized beds of substantial size have also been studied for the reaction of a solid, as in lime-burning and the roasting of ores, and the removal of sulfur oxides from stack gases. A remarkably uniform temperature can be maintained throughout the entire reactor, because of the high turbulence, the high heat capacity of the solid bed relative to the gas in it, and the high heat transfer rates between gas and solid associated with the large interfacial area. This uniform temperature enhances the selectivity that can be achieved. Other advantages are the continuous operation with few mechanically moving parts, the ability to add and remove solids easily, and the fact that more effective contact between gas and solid is obtainable than with competitive devices such as rotary kilns or tray reactors. The disadvantages are that some solids may agglomerate, or not flow freely, the attrition of the solid causes loss of material as fines which must be replaced, and dust removal equipment must be provided. The higher pressure drop relative to fixed-bed operation also increases costs, and the complex nature of the contacting causes uncertainties in scaleup.

As gas velocity is increased through a bed of finely divided solid particles, the pressure drop through the bed also increases until a point is reached at which the pressure drop just supports the weight of all the particles. At this point, known as incipient fluidization, particles begin to move with respect to one another. With further increase in gas velocity, the bed expands and ultimately particles present are blown out of the apparatus. As superficial gas velocity U is increased above the superficial fluidization velocity U_0, bubbles of gas appear in the fluidized bed, which grow in size as they rise. The fluidized bed may thus be regarded as a two-phase system: (a) a "particulate phase," or dense phase, in which the void fraction ε_0 is essentially that in the bed as a whole at incipient fluidization, and (b) "bubbles" containing little, if any, solid which carry all the additional gas beyond that required for incipient fluidization. The velocity at which gas flows through the dense phase corresponds approximately to that which produces incipient fluidization. As the bubbles rise they grow by seepage from the surrounding dense phase, the rate of which is a function of particle size and of particle-size distribution.

The gas velocity at incipient fluidization may be predicted by calculating the flow that produces a pressure drop through a fixed bed of particles equal to its weight per unit cross section. At incipient fluidization, the observed void fraction is about 0.4–0.5 and U_0 may be estimated as

$$U_0 = 0.00081(\rho_s - \rho)gd_p^2/\mu, \tag{2.29}$$

where ρ_s and ρ are the density of solid and gas (g/cm^3), d_p is the particle diameter (cm), and μ is the viscosity of the solid-free gas [89]. U_0 is the superficial velocity (cm/sec), volumetric flow rate per unit cross section of the bed. For a particle size of 200 μ diameter, typical values of U_0 are 3–6 cm/sec (0.1–0.2 ft/sec), and good fluidization is typically obtained at superficial velocities of three to four times this value. Many industrial operating conditions correspond to an over-all superficial velocity of 15–45 cm/sec (0.5–1.5 ft/sec). The maximum superficial velocity is limited by the excessive degree of entrainment and the increasingly poor contacting obtained.

Quantitative models for fluidized-bed reactors have been developed by many workers. Those of Davidson, Harrison, and co-workers [89] and of Rowe [279, 280] are particularly interesting because the results are analyzed in terms of the mechanics of bubbles in the fluidized bed, thereby providing a visualization of the processes occurring. Many uncertainties remain about the applicability of these and other models, industrial design remaining substantially empirical, but the main features are summarized below to provide a framework of understanding.

The bubbles are generally spherical but have an idented base that occupies about one-third of a complete sphere volume, so they are sometimes termed "spherical cap bubbles." Provided that the containing vessel is substantially larger than the bubble diameter, the rising velocity of the bubbles U_B (cm/sec) is about [89]

$$U_B = 0.79g^{1/2}V^{1/6} = 0.71g^{1/2}D_e^{1/2} = 22.2D_e^{1/2}, \tag{2.30}$$

where V is bubble volume and D_e is the bubble diameter. The rising velocity is essentially independent of particle size and particle density. For a typical bubble diameter of 7 cm, the rising velocity is about 60 cm/sec (2 ft/sec), some 10 to 20 times greater than the gas velocity in the dense phase. The bubbles are essentially empty, although from time to time a quantity of particles may descend from the roof and cause the bubble to split. The lowest one-third of a sphere enclosing the bubble is a wake of particles which travels upward with the bubble. Solid particles also flow in streamlines around the rising bubble so that a spout of material is also drawn up behind each bubble. The quantity of solids moving upward in the bottom of the sphere and in the spout below provides rapid mixing of solids from bottom to top.

The drag of particles flowing downward around the outside of the bubble sets up a recirculation pattern such that a concentric spherical "gas cloud" forms in the dense phase around the bubble and accompanies it upward. Gas flows upward through the bubble, down around the sides in a spherical shell in the dense phase around the bubble and re-enters the bubble at the bottom. There appears to be relatively little mixing between gas in this "gas cloud" and that in the remainder of the dense phase. The ratio of diameters of the spherical cloud to that of the bubble varies slightly with the ratio of bubble-gas velocity to that in the dense phase and is, for example, about 1.16, 1.1, and 1.05 for values of $U_B/U_0 = 2$, 10, and 20, respectively [279].

The superficial interstitial gas velocity is taken to be approximately equal to the superficial incipient fluidization velocity. From Equation 2.29, a typical value of the superficial interstitial gas velocity ($d_p = 200\ \mu$, viscosity for air of 0.02 cP) is about 3 cm/sec or about 1/20 of a typical bubble velocity. Since U_0 is proportional to the square of the particle size, the proportion of gas passing upward via the two paths is greatly affected by particle size and particle-size distribution. In a batch fluidized bed operated with a fixed charge of solid, attrition will gradually reduce particle size and thereby increase the fraction of the total gas passing upward in the form of bubbles.

As the bubbles rise, they grow in size by seepage of gas from the dense phase, which is at a higher pressure. The rate of growth increases with particle size because of the lower resistance of larger interstices to gas flow into the bubble. Bubbles are occasionally split by the descent of solids from the bubble roof, so there is an approximate maximum limit on their size. When bubble size approaches the diameter of the containing vessel, so-called "slugging" develops, in which the bubble fills the container cross section, a considerable fraction of the solid particles rain down through the gas slugs, and a different form of contacting develops. Typically, bubble diameters will grow to 1–2 in. within heights of 1–2 ft, so slugging frequently occurs in small laboratory-sized equipment. This can give a highly misleading prediction of performance to be expected in an industrial reactor which almost always is of such geometrical proportions as to operate in the bubbling rather than in the slugging mode.

The models of Davidson for analysis of reactions in fluidized beds assume that reaction occurs entirely in the particulate phase and that no reaction occurs within the bubble. The analysis centers around an estimation of the maximum bubble size that can exist, which is assumed to be constant through the bed. Interchange between gas in the bubble and that in the dense phase is assumed to occur by diffusion from the center of bubbles to the bubble wall and by "flowthrough" of gas in the particulate phase beside the bubble. Although quantitative methods are proposed for estimating the rates of mass transfer, these are still highly uncertain, particularly since there is little evidence as to the maximum bubble sizes which develop. For analysis of reaction data, the bubble diameter becomes the adjustable parameter. Typical values calculated from reaction studies are in the range of 5–10 cm, which is reasonably consistent with evidence from bed expansion studies, gas tracer studies, and observed velocities of bubble rise. However, with coarse particles bubble diameters of a foot or more may be encountered.

The model of Rowe assumes rapid interchange between the gas in the bubble and that in the gas cloud outside the bubble and in the wake comprising the bottom one-third of the bubble represented as a sphere, but interchange between the gas cloud and the remainder of the particulate phase is ignored. Thus, the gas leaving the top of the reactor is assumed to be a mixture of gases that have passed through the fluidized bed by two parallel paths, without interchange until the top is reached. On the contrary, interchange occurs continuously in the model of Davidson. Neither model provides a mechanism for back-mixing of gas in a vertical direction, although this does occur, particularly under vigorous agitation conditions. In the laboratory, a vapor is usually introduced through a porous disperser, so bubbles are initially very small and the best contact occurs over the first few centimeters of the bed. The performance of shallow beds with this type of disperser would be expected to be substantially better than that predicted by these models or from studies on deeper beds.

Direct experimental studies indicate that, with any appreciable degree of bubbling, excellent mixing of solids from top to bottom occurs. The solids are presumably accompanied by vapors in the interior of porous particles and possibly adsorbed as well. Rowe provides a specific method of estimating the rate of vertical mixing of solids. If a bubble is represented as a complete sphere, approximately the bottom one-third of the volume is filled with the dense phase. The rate of solids turnover is approximated by assuming that dense phase equal to one-third of the spherical volume of all the bubbles in the bed is carried from bottom to top of the bed at a velocity U_B. (Some of the particles in the bottom of the sphere drop out, but this is compensated for, at least in part, by material drawn up in the spout behind each bubble.) Typical conditions may correspond to a bed expansion (all due to bubble formation) of about 6 per cent, from which one would estimate that 2 per cent of the total solids are circulated from bottom to top at a velocity U_B. Taking 60 cm/sec for U_B, complete turnover in a bed height of 300 cm would occur in $(300/60)(1/0.02) = 250$ sec. If the average residence time of the solids is of the order of hours, the solid phase can be regarded as being well mixed. Data on vertical heat-transfer coefficients and on solids mixing suggest that the actual rate of solids mixing in a vertical direction is probably even greater than this, especially when using relatively high superficial velocities. Operation under slugging conditions would also presumably increase this rate. In both laboratory and industrial reactors the general pattern of solids recirculation is upward in the center and downward at the outside.

It will not be attempted here to analyze other models that have been proposed. Reference may be made to those of Lewis, Gilliland, and Glass [190], of Mathis and Watson [205], and of Lanneau [186], all of which like Rowe's assume the presence of catalyst associated with the bubble phase, in contrast with the models of Shen and Johnstone [316], of May [209], of van Deemter [342], and of Orcutt, Davidson, and Pigford [235], which, like that of Davidson and Harrison, do not. Heidel *et al.* [138] propose a two-zone model that divides the bed into a wall zone and a central or core zone. Several of these models provide for gas back-mixing. At gas velocities not greatly above that of incipient fluidization, the performance of a fluidized bed may closely approximate plug-type flow with respect to the gas, but at high fluidization velocities the performance may be even poorer than that calculated for a reactor with completely mixed gas [190]. For careful control of a gas-phase reaction in a fluidized bed it appears that optimum performance should occur at a gas flow rate providing a minimum number of bubbles, just sufficient for particle mixing and for desired rates of heat transfer from bed to wall, perhaps equivalent to about 3 or 4 × U_0. However, higher flow rates permit higher-capacity operation, and the nature of the contacting in practice is markedly affected by the particle-size distribution.

A great deal of information is available, largely in the patent literature, on the use of internals, such as horizontal and vertical screens, packing, etc., as a means of breaking up the bubbles and reducing back-mixing of gas. However, back-mixing of solids may also be reduced, which may then cause undesired temperature gradients in a vertical direction.

The particles in a fluidized bed are usually about 200–300 μ in diameter and smaller, but it is important that a substantial particle-size distribution exist and that a considerable portion of the particles, perhaps as much as 20 per cent or so, be below about 40 μ in size in order to avoid spouting and irregular behavior. Large industrial reactors typically have a height to diameter ratio of 2:1 or greater, in part to accommodate heat transfer equipment, cyclone separators, and other internals. In chemical processing it is also frequently desirable to introduce different reactants at substantially different locations in the reactor, which is facilitated by this kind of design.

Tremendous effort has been devoted to developing useful models for fluidized-bed reactors, but the performance of an industrial reactor is usually dominated by the largely empirical design and operation of the gas distribution grid. This grid is frequently an array of drilled holes or nozzles in a flat or dished plate; or it may be a spider or other form of distributor equipped with holes or nozzles to direct the incoming vapor sideways or downward. Finely porous distributors are seldom used industrially. Clearly the solids recirculation pattern will be markedly affected by distributor design. To avoid channeling and to achieve uniform gas distribution and stable fluidization, the pressure drop through the distributor must comprise a substantial fraction of the total through the reactor, perhaps of the order of 10 to 30 per cent, or even more for relatively shallow beds. The gas velocity from each aperture is consequently very high, typically exceeding 100 ft/sec and the jet may readily penetrate 6–12 in. or more into the bed before it begins to break up into bubbles. Particularly with relatively shallow beds, much of the reaction may occur in this highly turbulent regime rather than subsequently in the rising bubbles. Such high turbulence is, however, not altogether desirable since the mode of introduction of vapor may markedly affect the rate of attrition of some catalysts.

It is very difficult to use fluidized beds to achieve contact times less than 1 or 2 sec and about 10 sec is more typical. Short contact times can be achieved by carrying out most of the reaction in a transport line in which solid and gas move cocurrently. The design and operation of many fluid-bed industrial catalytic cracking units is aimed specifically to achieve this. For processing of chemicals, however, fixed beds even of the manifolded multitube variety are usually the design of choice to achieve controlled short residence times of the order of a second or less. Fluidized-bed reactors are treated in detail in the books by Davidson and Harrison [89] and by Kunii and Levenspiel [180].

2.5.1 *Mass Transfer*

Mass transfer in fluidized beds is excellent, partly because of the turbulent mixing, but mostly because of the large solid surface per unit volume. Richardson and Bakhtiar [268], for example, report that equilibrium between fluidized solids and effluent gas is very nearly reached when air containing toluene is passed at 1.3–6.0 cm/sec through only 0.2–7.0 cm bed depth of a fluidized granular toluene adsorbent consisting mostly of 40–80-μ particles. This means that mass transfer coefficients for fine particles are very difficult to measure (most of the data are for particles somewhat larger than those usually employed with gases, though in the range useful with liquids).

Correlations of existing data on mass transfer in both fixed and fluid beds are given by several writers [73, 103, 169, 266, 269, 315, 394]: Two are shown in Figure 2.1. Line *A-A* is that of Chu, Kalil, and Wetterath [73], representing data from various sources on gas-solid and liquid-solid systems in fixed and fluidized beds. A later correlation by Riccetti and Thodos [266], using the same coordinates, agrees closely with line *A-A*. A more recent correlation, by Sen Gupta and Thodos [315], is represented by the line *B-B*. Here A'_p is the surface area of a single particle, ε is the void fraction of the bed, and f is an "area availability factor."

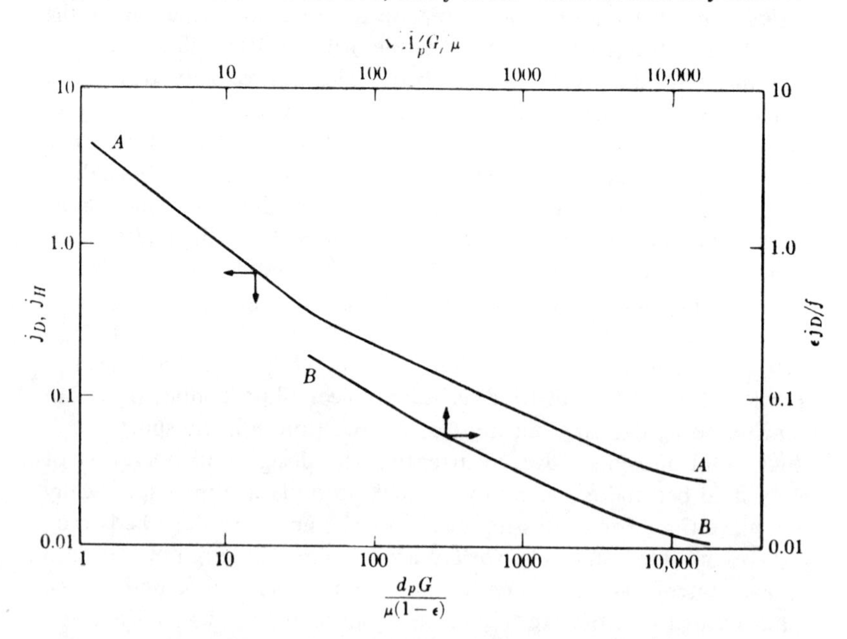

Figure 2.1. Correlation of data on mass and heat transfer in fluidized beds.

This last is 1.00 for spheres and varies from 0.82 to 1.36 for other shapes. Both lines in Figure 2.1 are based on extensive data for gas-solid and liquid-solid systems in both fixed and fluidized beds, and may be used to obtain good estimates of j_D and k_c for many operating conditions. The coordinates are dimensionless; either English or cgs units may be used.

The data of Fan, Yang, and Wen [103], for expanded or partly fluidized beds of 8-24-mesh benzoic acid particles in water fall considerably below line *A-A*, suggesting the possibility that Figure 2.1 may predict too large values of j_D for "expanded" beds.

Fluidized beds employing liquids are referred to as "slurry reactors"; these are discussed more fully in a later section.

Example 2.4 Estimation of Mass Transfer Rate in a Fluidized Bed

A fluidized bed is employed for the vapor-phase isomerization of compound A to form compound B, the reaction at the surface being first order in A, and irreversible. For operation at the specified conditions, show that the resistance to mass transfer from gas to particle is completely negligible. (The conditions described correspond to the production of 6 million pounds of product per year in a bed 3 ft in diameter.)

Data. Pressure, 1.0 atm; temperature, 810°K (998°F); feed, 20 mole per cent A in nitrogen; molecular weight of A, 100; superficial gas velocity, 50 cm/sec; gas viscosity, 0.0004 P; Schmidt number, 4.0; catalyst density, 3.0 g/cm³; particle diameter, 63 μ; void fraction in bed, 0.75; bed height, 100 cm; conversion of A to B, 90 per cent.

Calculation. The vapor feed rate is

$$G_M = \left(\frac{50}{22\ 400}\right)\left(\frac{273}{810}\right) = 0.000752 \text{ g-mol/sec} \cdot \text{cm}^2.$$

The average molecular weight of the gas mixture is 42.4, so the mass velocity is 42.4×0.000752, or 0.0319 g/sec · cm².

The catalyst surface is $6/d_p \rho_p = 6/0.0063 \times 3.0 = 318$ cm²/g, or $318 \times 100 \times (1 - 0.75) \times 3.0 = 23\ 800$ cm² above each cm² of reactor base.

The abscissa of Figure 2.1 is

$$\frac{d_p G}{\mu(1-\varepsilon)} = \frac{0.0063 \times 0.0319}{0.0004(1-0.75)} = 2.01,$$

whence j_D is 3.3. Then from the definition of j_D, given as Equation 2.9,

$$k_G = \frac{j_D G_M}{P N_{Sc}^{2/3}} = \frac{3.3 \times 0.000752}{1.0 \times 4^{2/3}} = 0.00099 \text{ g-mol/sec} \cdot \text{cm}^2 \cdot \text{atm}.$$

The rate of transfer of reactant from gas stream to catalyst surface is

$$\frac{0.000752 \times 0.20 \times 0.90}{23\,800} = 5.7 \times 10^{-9} \text{ g-mol/sec} \cdot \text{cm}^2.$$

The partial pressure difference necessary to cause transport at this rate is

$$\frac{5.7 \times 10^{-9}}{0.00099} = 5.66 \times 10^{-6} \text{ atm}.$$

The available reactant potential in the effluent is 0.020 atm, which is the partial pressure of A; the mean for the entire reactor is approximately this, depending on the nature of the contacting.

It is evident that the resistance to mass transfer to the catalyst particles is negligible, since transport requires only 0.03 per cent of the available potential. The gas immediately in contact with the solid surface is of very nearly the same composition as the gas stream.

2.5.2 *Heat Transfer*

As for mass transfer, the rates of heat transfer in fluidized beds are large, and the temperatures of the fluid and of the particle surfaces are very nearly the same. Line *A-A* of Figure 2.1 provides a good estimate of j_H as well as of j_D.

Experimental data of Richardson and Ayers [267] are in approximate agreement with those of Kettenring, Manderfield, and Smith [169], who suggest the following empirical equation for gas-fluidized systems:

$$\frac{hd_p}{k} = 0.0135\, N_{\text{Re}}^{1.3}. \tag{2.31}$$

where k is the thermal conductivity of the gas. This expresses their results for 20–40-mesh silica and alumina gel particles fluidized in air at values of N_{Re} ranging from 9 to 55. The definition of N_{Re} is the same as that used for fixed beds, namely, $d_p G/\mu$. Equation 2.31 may also be expressed in the form

$$j_H = 0.0135\, (N_{\text{Re}}/N_{\text{Pr}})^{1/3}. \tag{2.32}$$

Fridland [112] presents a recent summary and correlations of literature data on mass transfer in fluidized systems. Heat transfer coefficients from fluid bed to reactor wall in the presence of bubbling are one to two orders of magnitude greater than those observed in packed beds, presumably because of the rapid movement of particles to and from the wall.

2.6 Solid Catalysts Suspended in Liquids (Slurries)

Catalytic reactions may be carried out in a liquid reactant or a liquid in which the reactants are dissolved, the solid catalyst being suspended in granular or powdered form. The use of slurries is common in the chemical industry, particularly for hydrogenation. Batch operation is employed in units of small capacity, the reactor being a stirred autoclave.

Use of slurries for continuous operation would appear to offer many advantages over fixed beds, particularly for large-scale petroleum and petrochemical processes, where their use is not presently widespread. A stirred slurry can be kept at a uniform temperature throughout, eliminating the "hot spots" that detract seriously from the selectivity of catalysts in many vapor-phase fixed-bed operations. The large mass of liquid provides a safety factor in the case of an exothermic reaction which may cause a "runaway." Heat recovery is practical, since liquid-phase heat transfer coefficients are large. Pelleting costs may be avoided, and catalysts that are difficult to pelletize may be used. The small particles used in slurries may make it possible to achieve much higher rates of reaction per unit weight of catalyst than can be achieved with the large pellets employed in trickle beds if conditions are such that a low effectiveness factor would be encountered in the latter case. Catalysts can be regenerated continuously by removing a small part of the slurry, separating, regenerating, and returning the catalyst to the reactor. In many cases, the reactor can be a simple vessel in which the slurry is agitated by the gas fed. Alternately a stirred autoclave may be used, or an external pump may recirculate liquid slurry through an external heat exchanger and simultaneously provide agitation and maintain slurry suspension. Reactors can be staged by placing one above another in a single shell, the assembly operating very much like a bubble-tray rectifying column.

Slurry reactors may be used in two quite different ways. In the Fischer-Tropsch synthesis, carbon monoxide and hydrogen may be bubbled into a slurry of iron catalyst, forming hydrocarbons and oxygenated compounds that vaporize and are removed. The oil suspending the catalyst does not enter into the reaction, although it may be one of the products. Alternatively, the liquid may be the feed to the process, as in the hydrogenation of an unsaturated hydrocarbon.

One deterrent to the adoption of continuous slurry reactors is the fact that published data are inadequate for design purposes. It is not clear, for example, what designs would avoid possible plugging difficulties, and the data on capacity (allowable space velocity) are very limited. For many purposes, particularly oxidations, it is difficult to select a carrier liquid in which the reactants are soluble and which is stable at elevated temperatures in contact with the reactants and products. The behavior of a catalyst in contact with a liquid is possibly different than when in contact with a gas containing reactants with similar activities.

The ratio of liquid to catalyst in a slurry reactor is much greater than that in a trickle bed, and, hence, the relative rate of homogeneous reaction, if any, to catalytic reaction will be greater. The residence time distribution in a single slurry reactor is that characteristic of a well-stirred vessel in contrast to the close approximation to plug-type flow encountered in most industrial trickle-bed reactors.

2.6.1 *Mass Transfer in Slurry Reactors*

Reactants and products may be gases or liquids, and a variety of situations may exist, each requiring a somewhat different analysis of the mass transfer phenomena involved. As a typical example, the discussion that follows pertains to the hydrogenation of a pure liquid in which the catalyst is suspended. It will be assumed that the reaction is first order and irreversible, and that it occurs entirely on the external surface of the suspended particles. Hydrogen is bubbled into the slurry, absorption and reaction occurring simultaneously in the same reactor.

Hydrogen must be absorbed from the gas bubbles and be transported by liquid mixing and diffusion to the surface of the catalyst particles. In many instances the "gas side" resistance is negligible, and the liquid at the gas-liquid interface is essentially in equilibrium with the gas. This equilibrium concentration of hydrogen in the liquid will be designated c_e. The main body of the liquid is at a lower hydrogen concentration c_L. The concentration at the surface of the particle is c_s (g-mol H_2/cm^3). At steady state, the rate equations may be written as

$$\begin{aligned} N_v &= k_L A_B(c_e - c_L) \qquad \text{(gas interface to bulk liquid)} \\ &= k_c A_p(c_L - c_s) \qquad \text{(bulk liquid to particle)} \\ &= k_s A_p c_s \qquad \text{(surface reaction).} \end{aligned} \tag{2.33}$$

The molar flux N_v, the bubble surface A_B, and the catalyst surface A_p are based on unit *volume* of slurry as expanded by the gas flow. The last equation expresses the rate in terms of a surface reaction-rate coefficient k_s; k_L and k_c are mass transfer coefficients (cm/sec).

Combining these relations to eliminate c_L and c_s, we have

$$\frac{c_e}{N_v} = \frac{1}{k_L A_B} + \frac{1}{k_c A_p} + \frac{1}{k_s A_p}. \tag{2.34}$$

If both gas bubbles and catalyst particles are spheres of diameters d_B and d_p, respectively, then

$$A_B = 6H/d_B \tag{2.35}$$

and

$$A_p = 6m/\rho_p d_p, \tag{2.36}$$

where H is the gas "holdup" in the liquid (cm^3/cm^3 of expanded slurry), and m is the catalyst "loading" (g catalyst/cm^3 of expanded slurry).

Combining Equations 2.34, 2.35, and 2.36, we obtain

$$\frac{c_e}{N_v} = \frac{d_B}{6k_L H} + \frac{\rho_p d_p}{6m}\left(\frac{1}{k_c} + \frac{1}{k_s}\right). \tag{2.37}$$

In a series of tests varying only catalyst loading, the reciprocal of the rate should be linear in $1/m$. This has been confirmed by many investigators, among the earliest being Davis, Thomson, and Crandall [92]. Their results for the hydrogenation of several olefins are shown in Figure 2.2. Note that if the catalyst loading is already large, additional catalyst may do little to speed up the reaction, i.e., the rate of absorption of hydrogen is controlling. Another example is shown in Figure 2.3, representing the data of Sherwood and Farkas [317] on the hydrogenation of α-methylstyrene to cumene in a 2.5-cm-i.d. bubble column, using a catalyst suspension of palladium black in the form of 55 μ particles. The data are for 1 atm pressure of hydrogen. This method of graphing is useful for correlation of kinetic data but it does not provide a means of determining the relative importance of mass transfer to catalyst particles and surface reaction (k_c versus k_s in Equation 2.37). For this we need an independent method of predicting k_c, which will be discussed below. At low catalyst levels of loading the rate of reaction may be determined primarily by the amount of impurities present in the system that poison the catalyst, in which case the correlation of data plotted in the manner of Figures 2.2 and 2.3 would curve upward at high values of $1/m$. At very high catalyst loadings (low values of $1/m$), k_c may be reduced and the lines flatten as the intercept is approached.

If hydrogen absorption is rapid, the liquid can be kept essentially saturated ($k_L A_B$ large, $c_e = c_L$). Then, if the catalyst is highly active ($k_s \gg k_c$), the over-all rate will be equal to the rate of mass transfer from liquid to catalyst particles, and the rate equation simplifies to

$$N_v = k_c A_p c_e = \frac{6mk_c c_e}{\rho_p d_p}. \tag{2.38}$$

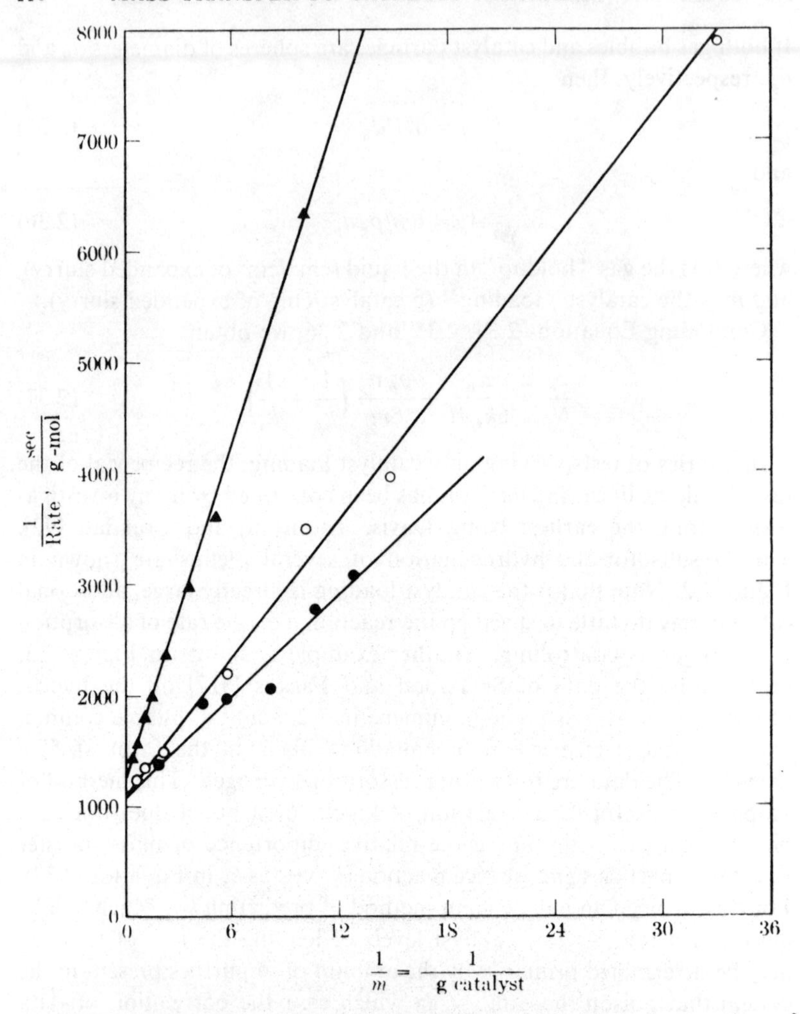

Figure 2.2. Effect of catalyst loading on rate of batch hydrogenation of 2-methyl-2-butene $(CH_3)_2C{=}CHCH_3$ (top line), 2-pentene, $CH_3CH_2CH{=}CHCH_3$ (second line), and 2-vinyl propane, $(CH_3)_2CHCH{=}CH_2$ (bottom line). Davis, Thomson, and Crandall [92].

As noted earlier, this is the maximum rate; lower rates will be observed if the catalyst is not sufficiently active, or if the liquid is not kept saturated with hydrogen.

When the catalyst loading is large ($1/m \rightarrow 0$), the rate becomes controlled by the rate of absorption of hydrogen and

$$N_v = \frac{6k_L H c_e}{d_B}. \tag{2.39}$$

The intercepts in Figure 2.2 or Figure 2.3 are thus inversely proportional to the gas absorption coefficient. Those determined by Sherwood and Farkas [317] in this fashion are of the same magnitude as those reported in the literature.

Calderbank *et al.* [57] showed the diffusion of hydrogen from gas bubbles to liquid to be the rate-controlling step in a study of the hydrogenation of ethylene at 40–60°C by bubbling the gas mixture through a suspension of Raney nickel catalyst in toluene or *n*-octane. Gas-liquid hydrogen diffusion was likewise at least a part of the resistance in a study of the Fischer-Tropsch reaction as carried out by

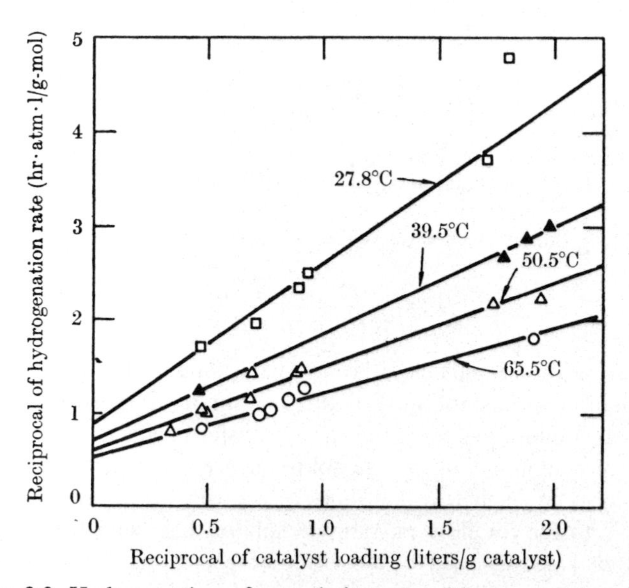

Figure 2.3. Hydrogenation of α-methylstyrene with suspended particles of palladium black in a bubble column; superficial hydrogen velocity = 2.3 cm/sec. Sherwood and Farkas [317].

bubbling synthesis gas through a slurry of iron catalyst suspended in molten wax. In this case a further complexity was introduced by the formation of free carbon which reduced the degree of gas-liquid contacting and caused catalyst fouling.

A recent review by Battino and Clever [23] presents a detailed guide to the literature on solubility of hydrogen and other gases in various liquids. It may be noted that the solubilities of gases having low critical temperatures usually *increase* with temperature, and may be surprisingly high. At 67 atm (1000 psia) the solubility of hydrogen in n-hexane at 204°C (400°F) corresponds to a mole fraction of 0.18.

2.6.2 *Mass Transfer to Bubbles and Suspended Solid Particles*

The mass transfer coefficient k_c for a single sphere at rest in a large volume of stagnant fluid is given by

$$k_c = 2D_{1m}/d_p, \tag{2.40}$$

where D_{1m} is the molecular diffusion coefficient for the species diffusing in the fluid. This equation is easily derived. Equate the diffusion rate at the sphere surface to that through a shell of radius r (N positive in the direction of increasing r):

$$N(\pi d_p^2) = -D_{1m}(4\pi r^2)\frac{dc}{dr},$$

$$N\frac{d_p^2}{4}\left(\frac{2}{d_p} - \frac{1}{r}\right) = D_{1m}(c_s - c).$$

Substituting the limits $c = c_L$ at $r = \infty$, we have

$$k_c = \frac{N}{(c_L - c_s)} = \frac{2D_{1m}}{d_p}.$$

Any motion of the particle relative to the fluid will tend to make k_c larger than the value obtained by this relation. (For most other shapes, such as an infinite cylinder, the mass transfer coefficient from a large volume of stagnant fluid to the solid surface never reaches a steady-state value; the coefficient continues to decrease with time.)

Mass transfer for flow past single spheres has been the subject of much study [258, 327]. As for most bluff objects, transition from laminar to turbulent flow is gradual, and attempts have been made to correlate the existing experimental data by means of single equations covering the entire range of Reynolds numbers. Several of the proposed correlations are of the form

$$\frac{k_c d_p}{D_{1m}} = 2.0 + K(N_{\mathrm{Re}})^{1/2}(N_{\mathrm{Sc}})^{1/3}. \tag{2.41}$$

Values of the constant K reported by various authors fall in the range 0.3–1.0. A recent review of existing data in the range of Reynolds numbers from 20 to 2000 by Rowe and Claxton [281] (the region of particular interest in catalysis), suggests $K = 0.63$ for gases and $K = 0.76$ for liquids.

Mass transfer between spheres and the surrounding fluid is subject to exact analysis for situations where the relative velocity of the two is low (as in slurries of small particles in a liquid). Several analyses have been published, perhaps the most complete being that of Brian and Hales [46]. The differential equations for diffusion and for the flow field around the sphere are combined and solved numerically, giving a relation between the Sherwood number, $k_c d_p/D_{1m}$, and the Peclet number, $d_p U/D_{1m}$. The result is shown by the curve on Figure 2.4.

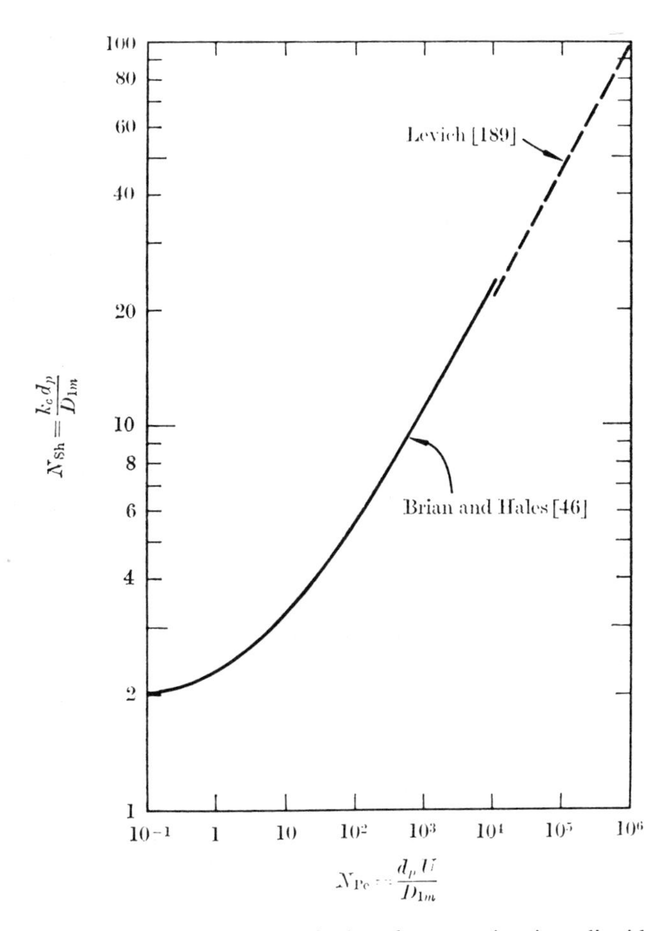

Figure 2.4. Mass transfer to a single sphere moving in a liquid at low velocity.

The intercept at $N_{Pe} = 0$ is 2, as given by Equation 2.40, and the upper dashed linear asymptote is Levich's [189] solution for $N_{Pe} > 1000$, represented by

$$\frac{k_c d_p}{D_{1m}} = 0.997 N_{Pe}^{1/3}. \tag{2.42}$$

Brian and Hales allowed for angular diffusion over the sphere surface, though Levich did not.

The solid curve of Figure 2.4 is well represented by the following equation

$$\left(\frac{k_c d_p}{D_{1m}}\right)^2 = 4.0 + 1.21 N_{Pe}^{2/3}. \tag{2.43}$$

Slurry reactors employ solid catalysts in the form of small particles suspended in a fluid. Except in the case of ion exchange resins the particles are seldom spherical. The relative velocity of particle and fluid is roughly that of free fall due to gravity but the mass transfer is enhanced by the turbulence induced both by mechanical agitation and by rising gas bubbles.

It has proved difficult to characterize the shapes of broken irregular solid particles in a simple way so as to relate their dimensions to the diameter of a sphere in order to employ the correlations of data on mass transfer to single spheres. The paper by Lochiel and Calderbank [191] is an important contribution to this problem, but it is often adequate in technical calculations to employ a diameter corresponding to a sphere having the same ratio of external surface to volume (or settling velocity) as the irregular particle.

The terminal velocity of small spheres is given by Stokes' law:

$$U_{\text{term}} = \frac{g d_p^2 \,\Delta\rho}{18\mu}. \tag{2.44}$$

The corresponding Peclet number, denoted by N_{Pe}^*, is

$$N_{Pe}^* = \frac{g d_p^3 \,\Delta\rho}{18\mu D_{1m}} \quad \text{(free fall)}. \tag{2.45}$$

Here g is the acceleration due to gravity, $\Delta\rho$ is the difference in density between particle and fluid, and μ is the fluid viscosity.

Equation 2.45 may be employed with Figure 2.4 to obtain k_c for a sphere settling at its terminal velocity. For the Levich asymptote this leads to

$$k_c^*(N_{Sc})^{2/3} = 0.38\left(\frac{g\mu\,\Delta\rho}{\rho_L^2}\right)^{1/3} \tag{2.46}$$

Here k_c^* is the value of k_c for a sphere settling at its terminal velocity, and ρ_L is the density of the fluid. This result agrees with Calderbank and Jones' data [58] on small resin beads and gas bubbles to within about 15 per cent.

Harriott [133], following a similar approach, employed Stokes' law to eliminate velocity from Ranz and Marshall's correlation [258] of k_c in the form of Equation 2.41 for flow past single spheres. At low settling velocities ($N_{Pe} < 500$), k_c^* is best estimated by combining Equation 2.45 with Figure 2.4, as suggested above. Figure 2.5 shows the relation

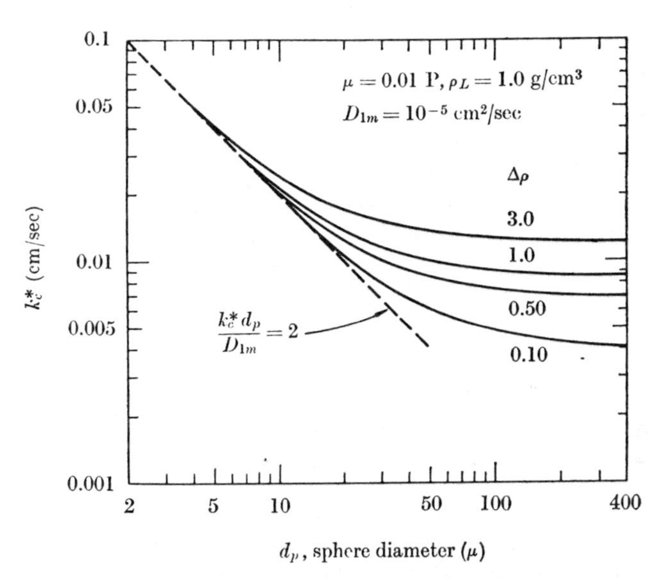

Figure 2.5. Calculated mass transfer coefficient for a single sphere settling at its terminal velocity in a liquid.

between k_c^* and sphere diameter calculated in this way for values of μ, ρ_L, and D_{1m} typical of solids settling in a liquid.

Slurries of catalyst particles suspended in a liquid, as used in catalysis, are necessarily agitated vigorously both to keep the solids in suspension and to promote absorption of reacting gases. The turbulence that results promotes mass transfer, and the actual value of k_c is greater than k_c^*. The difference is not as great as might be anticipated, however, since the particles tend to follow the motion of the fluid. Harriott's comparison of his own and published data on such systems showed the ratio k_c/k_c^* to fall generally between 1 and 4. As a first approximation k_c may be estimated to be twice the value calculated by the use of Figure 2.4, with N_{Pe} based on the terminal velocity of the particles in free fall (Equation 2.45).

The correlation of Brian and Hales [46] for mass and heat transfer to spheres suspended in an agitated liquid relates k_c to the power input to the slurry. This provides what appears to be the soundest method of estimating k_c, provided that the power input is known or can be calculated with reasonable assurance. Figure 2.6 shows the result of Brian and

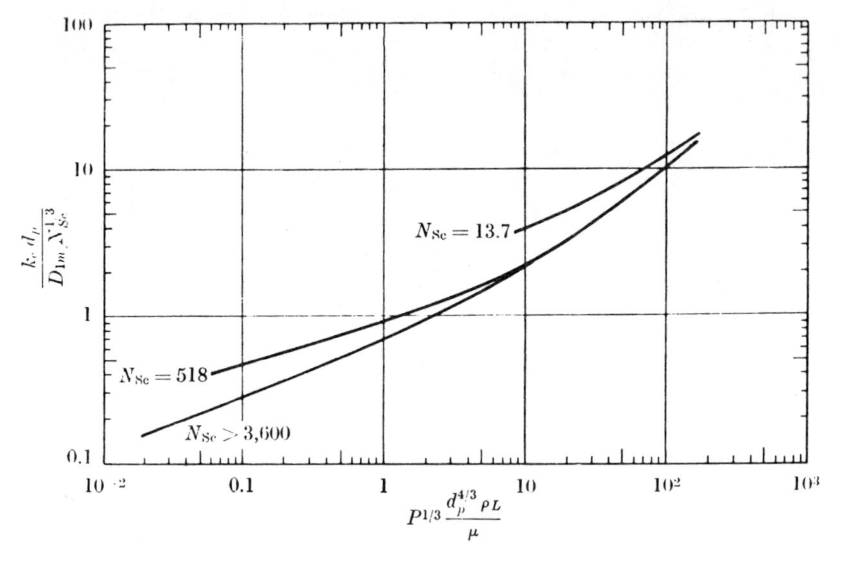

Figure 2.6. Effect of power input on mass transfer to spheres suspended in an agitated liquid. Brian and Hales [46].

Hales' analysis, which was found to be well supported by a variety of data. The ordinate is $(k_c d_p/D_{1m})/N_{Sc}^{1/3}$. The abscissa, $P^{1/3}d_p^{4/3}\rho_L/\mu$, involves the power input P expressed as power per unit *mass* of slurry. Both groups are dimensionless; in cgs units P must be expressed in ergs/sec · g slurry (1 kW is 1000 J/sec, or 10^{10} ergs/sec). Rushton, Costich, and Everett [286] and O'Connell and Mack [229] provide useful correlations of agitator power as a function of vessel geometry and stirrer speeds for turbine and propeller-type agitators; these may be used to estimate P.

Brian and Hales extended their analysis to show the effects of changing particle size (as in the dissolution of a soluble salt) and of transpiration velocity (the radial diffusion flux) on k_c. These effects would not appear to be important in heterogeneous catalysis, since the catalyst particle does not change in size, and the diffusion flux is ordinarily slow.

Calderbank [56] describes methods of estimating the minimum agitation intensity required to maintain the catalyst particles in suspension.

Measurements of reaction rates using catalysts suspended in liquids can provide mass transfer data if conditions are such that the rate is controlled by diffusion to the catalyst particles. One study in which this condition was apparently met is that of Polejes, Hougen, and co-workers [254]. A 300-mesh suspension of 2 per cent palladium on alumina was used to hydrogenate α-methylstyrene (to cumene), with pressure operation (1.3 to 3.7 atm) in a 1-gal stirred autoclave. At low stirrer speeds, the resistance to hydrogen absorption was appreciable; but at $N_{Re} = 100\ 000$, $1/k_L A_B$ evidently became negligible. In this case $N_{Re} = N_s d_I \rho/\mu$, where N_s is the stirrer speed (rps), and d_I is the diameter of the impeller.

Figure 2.7 shows Polejes' data at the highest stirrer speeds, plotted

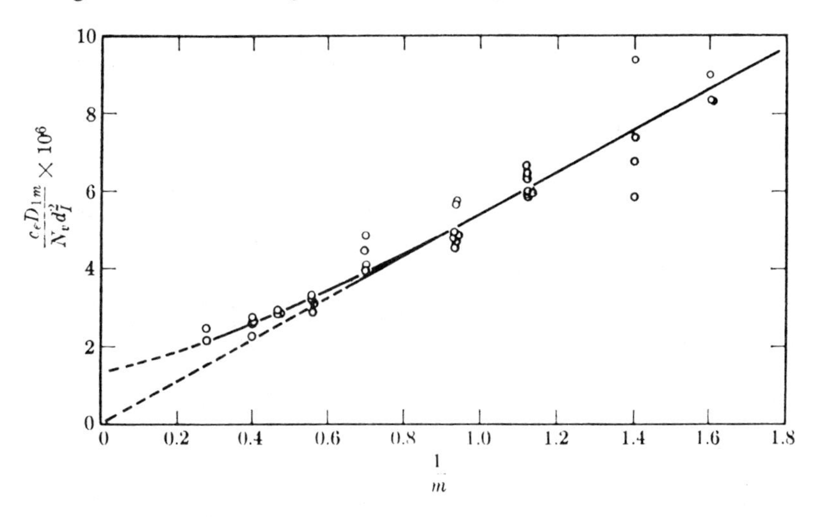

Figure 2.7. Effect of catalyst loading on the rate of hydrogenation of α-methylstyrene. Polejes *et al.* [254].

versus $1/m$ as suggested by Equation 2.37; d_I and D_{1m} are introduced to define a dimensionless "resistance." A line through the origin represents the data well at low catalyst loadings; the intercept at $m = \infty$, indicated by the curved line through the data points, is perhaps due to settling or agglomeration of catalyst particles at very high catalyst loadings rather than to a resistance to hydrogen absorption from bubbles.

The catalyst particles were said to have a surface mean diameter of 7.15 μ, a volume mean diameter of 13.4 μ, and a specific gravity of 4.16 g/cm^3. This corresponds to an external particle surface of $304m$ cm^2/l. Using this value for A_p, one can easily show that the straight line through the origin of Figure 2.7 represents a single value of $k_c d_p / D_{1m}$ of 5.4. (The k_c calculated in this way is actually the reciprocal of the quantity in parentheses in Equation 2.37; k_s is assumed to be large.) The Peclet number based on the free-settling velocity was 0.13, so $k_c^* d_p / D_{1m}$ is 2.0 (Figure 2.4). Considering the fact that the particles were of irregular shape and not of uniform size, the agreement between k_c from the hydrogenation data and k_c^* from the diffusion theory is surprisingly good. The ratio k_c/k_c^* of $5.4/2.0 = 2.7$ is in the range found by Harriott.

Since k_c and particle surface per unit mass are each inversely proportional to the particle diameter, the use of a slurry can effectively eliminate the resistance to diffusion between the ambient fluid and the outer surface of the catalyst particle. What resistance there is can usually be estimated with adequate assurance by the methods described above. Diffusion and reaction within the pores of the solid catalyst are discussed in Chapters 3 and 4. The resistance to transport of the reacting gas from gas bubbles to ambient liquid is discussed briefly below.

2.6.3 *Mass Transfer from Bubbles*

The extensive literature on mass transfer from gas bubbles to liquids will not be treated here, since an excellent review has been published by Calderbank [56] and detailed treatments are available in the books by Sherwood and Pigford [319], by Valentin [341], and by Hobler [152]. A few general comments, however, may help provide orientation. Swarms of small bubbles (e.g., <2.5 mm diameter) behave like rigid spheres, and mass transfer coefficients from the interface to the liquid are correlated by the same relationship that applies for small solid spheres. Mass transfer coefficients from larger bubbles are slightly greater since the bubbles undergo internal circulation and their shape becomes distorted as they move through the liquid. A correlation [56] for these larger bubbles represents the results of a wide variety of studies.

$$k_L (N_{Sc})^{1/2} = 0.42 \left(\frac{\mu g \, \Delta\rho}{\rho_L^2} \right)^{1/3} \tag{2.47}$$

As in the previous section, N_{Sc} is the Schmidt number for the *liquid*.

Gas bubble size is affected by many variables, and Calderbank [56] gives guidance on methods of prediction. At the point of introduction of the gas into the liquid the bubble size is markedly affected by such variables as orifice diameter and gas velocity in the orifice. However, coalescence occurs as bubbles rise and their size may vary substantially with distance from the disperser. The high shear at the point of introduction of the gas probably substantially increases the local mass transfer coefficient, but in a deep reactor vessel this zone may represent only a small fraction of the total interfacial area in the vessel. In a mechanically agitated vessel the degree of power dissipation affects the average bubble size, but has little effect on the mass transfer coefficient as such. Small amounts of surface-active agents, such as are often encountered in practice, or solutions of electrolytes, may substantially increase the total interfacial area. For gas dispersions in a pure liquid of relatively low viscosity as produced on the sieve plates used in distillation or gas absorption, Calderbank suggests as a first approximation for high gas flow rates that the mean bubble diameter may be taken as 4.5 mm, with a corresponding gas holdup of 60 per cent and interfacial area of 8 cm^2/cm^3 of dispersion. Chemical reactors will usually be much deeper than sieve plates and somewhat lower interfacial areas would be expected. Industrially, such reactors are frequently operated at as high gas velocities as possible for maximum capacity. At atmospheric pressure a superficial gas velocity of the order of about 0.5 ft/sec (based on total cross-sectional area of the reactor) represents an approximate upper limit above which entrainment may become excessive. The corresponding gas holdup is about 0.20. For deep vessels not mechanically agitated and liquids of relatively low viscosity (of the order of 1 cP) mean bubble diameters will be in the range of about 0.3–0.6 cm [56, 156]. They will be larger for more viscous liquids. Hughmark [156] presents a correlation for gas holdup in bubble columns which, assuming $d_B = 0.6$ cm, predicts, for example, an interfacial area of 1.3 cm^2/cm^3 at a superficial velocity of 0.25 ft/sec and nearly double that at 0.5 ft/sec. Direct measurements by Calderbank *et al.* [57] of interfacial area in a 2-in.-diameter column when feeding synthesis gas into molten wax at 265°C ($\mu = 1.6$ cP) gave $a = 1.5$ cm^2/cm^3 at a superficial velocity of 0.05 ft/sec and about 4 cm^2/cm^3 at velocities of 0.12–0.18 ft/sec.

In mechanically agitated vessels d_B is also affected by the mechanical design of the impeller, the geometry of the tank, and rotation speed; the total interfacial area is substantially increased if the agitator is operated at such high speeds that gas is sucked down into the liquid by vortexing. The residence time distribution of the gas in a vessel closely approximates plug flow at low superficial gas velocities but at high degrees of agitation it may approach complete mixing, in which case the composition of the gas phase throughout the reactor is the same as that of the gas leaving the vessel.

Kozinski and King [175] recently presented a summary of liquid-phase mass transfer coefficients reported in the literature for the unbroken top interface in a stirred vessel as well as new data of their own. Information is also presented of the effect on $k_L a$ as stirring speeds are increased to cause a vortexing action to develop.

Diffusion limitations in a slurry reactor can have a profound effect upon the selectivity of a complex reaction. This is illustrated in the hydrogenation of fatty oils using a suspended nickel catalyst, a process which has been of technical importance for many decades and which is usually carried out by batch processing. The oils contain a variety of saturated and unsaturated fatty acid groups and it is generally desired to achieve a degree of selectivity corresponding to hydrogenation of multiple double bonds in the fatty acid group to about one double bond per group (e.g., oleic acid) with minimum formation of saturated fatty acids, (e.g., stearic acid). An unusual feature of this reaction is that maximum selectivity is obtained by operating under highly diffusion-limited conditions with respect to hydrogen. This is achieved industrially by using low hydrogen pressure, low stirring intensity, high catalyst concentrations, and high temperature, which result in a low hydrogen concentration on the catalyst surface. In polyunsaturated fatty acids, conjugated double bonds are hydrogenated much faster than isolated double bonds. The low hydrogen concentration on the catalyst is apparently desired because it accelerates the isomerization rate of multiple double bonds to the conjugated form, relative to the rate of the hydrogenation reaction, and the conjugated double-bond structures can then be hydrogenated with high selectivity. An additional complication is that the desired softening characteristics of the product with temperature change are improved by a relatively low concentration of *trans* isomers of the partially hydrogenated oil, but the rate of isomerization of *cis* to *trans* isomers relative to the rate of hydrogenation is also enhanced by low hydrogen concentrations [396].

A variety of diffusion effects probably occurs industrially without being recognized in many hydrogenation and other reactions carried out in slurry reactors. This can be an important cause for the difference in behavior between laboratory and plant-size equipment. Even with the fine particle sizes characteristic of slurry-type operation, diffusion limitations within a porous catalyst may be significant; i.e., the effectiveness factor may be substantially less than unity with consequent loss in selectivity. An example, again in fatty oil hydrogenation, is given by Coenen [75]. With three supported nickel catalysts, all of about 6–7 μ particle size, the selectivity was much poorer with one having an average pore diameter of 27 Å than with the other two having average diameters of 34 and 66 Å. Coenen estimates the diameter of the triglyceride molecule to be about 15 Å, calculated as a sphere, which is slightly over one-half of the average channel size. Evidently a highly restricted rate of diffusion, like that encountered in molecular sieves, was occurring.

Considering batch-type slurry reactor processing in general, it is evident that if mass transfer from gas bubbles to the liquid is controlling, the reaction will appear to be zero order. This condition is most likely during the initial portion of the reaction and will persist until the reactant concentration drops below some critical range. Stirring intensity affects gas-liquid transport more than liquid-solid transport, since it increases gas-liquid interfacial area; hence, the zero-order region should persist for a shorter length of time with high-intensity mixing. The technique of increasing the stirring rate until no further increase in reaction rate occurs may usually be taken as a criterion for absence of mass transfer limitations, in either batch or continuous processing, if the range of power input studied is sufficiently wide that a significant variation in k_c would be expected from Figure 2.6. Above a certain speed of rotation an impeller may rotate in a vortex in such a manner as to produce little further mixing of liquid and gas, particularly if the vessel is not baffled. A variety of reactors may be used in the laboratory for hydrogenation reactions, differing greatly in the degree of agitation and contacting that they provide.

The above analyses have several significant implications for the planning of experiments and interpretation of results. If the concentration of dissolved hydrogen in the liquid next to the catalyst surface is too low, the hydrogen concentration adsorbed onto the surface of a catalyst is also low and so-called "hydrogen-deficient operation" takes place. Partially hydrogenated products then frequently accumulate on the surface of the catalyst and deactivate it by undergoing side reactions such as polymerization and condensation. If a hydrogen-rich catalyst is desired, it is seen that vigorous agitation, low temperatures, high hydrogen pressure, low catalyst slurry concentrations, and a partially deactivated catalyst are all more or less equivalent. Some desiderata are incompatible, such as that of high capacity and low catalyst concentrations, but the investigator alert to the possible effects of mass transfer can thread his way through the variables at his command to an appropriate optimization. Agglomeration of the catalyst will usually decrease the rate of reaction markedly, but dispersal may be enhanced by changing the nature of the catalyst carrier, the *p*H, or the nature of the solvent, although the intrinsic kinetics of the reaction as such may also be greatly affected by changing the liquid environment. Sokol'skii [326] reviews the results of a substantial number of hydrogenation studies in which diffusion effects played a role.

It is evident that the existence of substantial mass transfer limitations between gas bubbles and liquid is equivalent to operating a reactor with intrinsic kinetics controlling but at a lower gas pressure. Hence, the effect of this type of mass transfer limitation on reaction selectivity is equivalent to the effect of hydrogen concentration in the liquid on intrinsic selectivity. If mass transfer from liquid to catalyst is the controlling step, however, the observed selectivity depends upon the diffusion characteristics of products and reactants.

If adequate provision for gas absorption is not provided, as in the not-uncommon practice of simply admitting hydrogen over the surface of the liquid in a stirred autoclave used for hydrogenation, the rate may be limited almost entirely by the possible rate of hydrogen absorption. In this situation it does little good to add more catalyst. The intercept on Figure 2.3 is a large fraction of the total value of the ordinate, which is little affected by increasing m. The reaction rate per unit volume in this situation is roughly proportional to the ratio of the gas-liquid interfacial area to the liquid volume.

Two other diffusion processes may need to be considered in some cases. The potential rate of diffusion of liquid reactant to the catalyst surface will generally be much greater than that of hydrogen because it is usually present in much higher concentration; but at high degrees of conversion, as toward the end of a batch process, this might become a significant limitation on an active catalyst. Whether diffusion of product from catalyst surface to bulk liquid is a significant limitation depends greatly upon the relative adsorptivity of products and reactants on the catalyst. In many hydrogenation reactions the reactant is an unsaturated compound such as an olefin, aromatic, or oxygenated material which is more strongly adsorbed than a saturated hydrocarbon product so that product diffusivity is unimportant. However, the opposite would frequently be the case in dehydrogenation or possibly in some hydrodesulfurization reactions in which the H_2S formed may be more strongly adsorbed than an organo-sulfur compound. Here the relative concentrations of adsorbed species and, hence, the course of the reaction will be markedly affected by the product concentration in the liquid in contact with the catalyst surface. This, in turn, may be affected by the mass transfer characteristics of the product away from the catalyst.

2.7 Hydrogenation Capacity of Slurries

It is of interest to estimate the hydrogenation capacity of continuous slurry reactors and to express the rates in terms of space velocities attained in gas-phase fixed-bed reactors. When the hydrogen "uptake" is of the order of 89 m^3/m^3 liquid feed (500 SCF/bbl*), many fixed-bed hydrogenations are operated with liquid hourly space velocities of 0.5 to 2.5.

Koppers [173, 174] described the results of tests on a slurry reactor for the production of light hydrocarbons from carbon monoxide and hydrogen. The two gases were bubbled into 10 m^3 (354 ft^3) of high-boiling paraffin oil in which 880 kg (1940 lb) of powdered iron catalyst was suspended. Products were vaporized and were recovered from the effluent gas. At 268°C (514°F) and 12 atm, the hydrogen reacting amounted to 4000 g-mol/hr · m^3 slurry. This corresponds to an LHSV of 1.1 volumes per hour per volume of slurry for a process in which a continuous liquid feed is hydrogenated to the extent of 89 m^3/m^3 liquid. Assuming the rate of hydrogenation to be proportional to the hydrogen pressure (which was only 4.8 atm in the Koppers tests) the LHSV would be 11 at 48 atm (705 psia) hydrogen pressure. The rate would be still greater at the somewhat higher temperatures often employed in hydrogenations, since the solubility of hydrogen increases with temperature.

A patent by Shingu [320] describes the use of a heavy slurry of silver oxide on silica gel in liquid dibutyl phthalate for the oxidation of olefins. It is claimed that conversions of 15 per cent and selective yields of 91 per cent can be obtained in the oxidation of ethylene to ethylene oxide at 0.8–5 atm and 220°C, using a feed mixture containing 20 per cent oxygen and a space velocity of 850 volumes feed per unit volume of slurry. If valid, this claim corresponds to a gas uptake of 5260 g-mol/hr · m^3 slurry, or 0.33 lb-mol/hr · ft^3 slurry, which is equivalent to an LHSV of 1.3 for the hydrogenation of a liquid feed to the extent of 89 m^3/m^3 of liquid. Providing that inert liquid carriers can be found, slurry catalysis would appear to have considerable advantages in oxidation processes, since the uniform catalyst temperatures tend to make the reaction more selective.

* SCF refers to cubic feet of gas at 60°F and 1.0 atm.

The "H-Oil" process [68, 69, 70, 309] employs a heavy slurry of catalyst particles in oil (e.g., 40 g of 1–0.5 mm granules per liter of slurry) for hydrocracking and hydrotreating various stocks, especially heavy residual oils containing metals. The solids content of the slurry is so great and the particles so large that the upflowing oil approximates plug flow, and catalyst is not carried overhead. With an LHSV of 30, sulfur and nitrogen removals of 90 and 30 per cent, respectively, were obtained at 76 atm (1100 psig) with a feed of furnace oil containing 1.0 per cent sulfur and 300 ppm nitrogen.

In the hydrocracking and hydrotreating of a Kuwait residuum at 204 atm (3000 psia), the hydrogen "uptake" was 267 m^3/m^3 of liquid feed (1500 SCF/bbl), when using LHSV's of 2.0 and 1.5 in the two sections of the reactor. These values were based on fresh feed; the recycle rate was high, so that the LHSV based on actual oil flow was 20 to 23. The same values of LHSV should be attainable at 1000 psia for a hydrogen consumption of 89 m^3/m^3 liquid (500 ft^3/bbl), provided that the catalyst was sufficiently active to make mass transfer controlling.

The three examples cited and the preceding theoretical analyses suggest that slurry reactors may have quite high capacities. A trickle-bed reactor can contain a substantially greater quantity of catalyst per unit volume of reactor than a slurry reactor, so it permits a greater capacity to be obtained under processing conditions corresponding to low rates of reaction. However, slurries appear to be particularly advantageous for use with highly active catalysts which, in particle sizes typical of packed-bed operation, make little effective use of their large internal surface.

For highly exothermic reaction in which essentially complete conversion is desired, a tandem continuous reactor system appears attractive — a slurry reactor to provide good temperature control during the main portion of the reaction, followed by a trickle bed to finish the reaction when heat transfer problems are relatively insignificant. Such a reactor system is used in the Institut Francais du Petrole process for production of high-purity cyclohexane by hydrogenation of benzene.

Example 2.5 Estimation of the Capacity (LHSV) of a Slurry Reactor for Hydrogenation

An unsaturated hydrocarbon oil is to be hydrogenated at 316°C (600°F) and 54.5 atm (800 psia) using a catalyst suspended as a slurry carrying 8.0 g catalyst per liter of oil. Assuming that the oil can be maintained saturated with hydrogen, what LHSV can be expected if the reaction consumes 89 m^3

hydrogen (15°C, 1 atm) per m³ of liquid feed (500 SCF/bbl)? Assume that the catalyst is very active, and that the rate of hydrogenation is controlled by the rate of mass transfer of hydrogen from liquid to catalyst particles.

Data. Molecular weight of oil = 170
Specific gravity = 0.51 at 316°C
Viscosity = 0.070 cP
K of hydrogen in oil = 5.0 (mole fraction in gas/mole fraction in liquid)
D_{1m}, hydrogen in oil at 600°F = 5.0×10^{-4} cm²/sec
Catalyst particle diameter = 10 μ (0.001 cm). These will be taken to be spheres.
Particle density = 3.0 g/cm³

The mass transfer coefficient k_c^* will be estimated from Figure 2.4:

$$N_{Pe}^* = \frac{d_p^3 g \Delta\rho}{18\mu D_{1m}} = \frac{(0.0010)^3 \times 981 \times (3.0 - 0.51)}{18 \times 0.00070 \times 5.0 \times 10^{-4}} = 0.39,$$

whence $k_c^* \, d_p/D_{1m} \sim 2$. k_c may be approximated as $2k_c^*$.

$$k_c = \frac{2 \times 2 \times 5 \times 10^{-4}}{0.0010} = 2 \text{ cm/sec.}$$

Particle surface $= \dfrac{6m}{\rho d_p} = \dfrac{6 \times 8.0 \times 10^{-3}}{3 \times 0.001} = 16$ cm²/cm³ slurry in absence of gas.

The hydrogen concentration in oil (taking mole fraction in gas to be unity) is 1/5.0 = 0.2 mole fraction H_2. The average molecular weight of the liquid is 136. Estimating the specific gravity of the hydrocarbon-rich oil to be 0.45, we find the hydrogen concentration to be 0.66 g-mol/l = 0.00066 g-mol/cm³. The hydrogen transfer rate is 2 × 16 × 1000 × 0.00066 = 21.2 g-mol/sec · l.

$$\text{LHSV} = \frac{21.2 \times 22.4 \times 288 \times 3600}{89 \times 273} = 20\,200 \text{ m}^3/\text{hr} \cdot \text{m}^3$$

$$= 20\,200 \text{ ft}^3/\text{hr} \cdot \text{ft}^3 \text{ bed.}$$

The result is, of course, extremely high. Diffusion coefficients for hydrogen in oils at elevated temperatures have not been measured; the value assumed is estimated from published values for methane in several hydrocarbons [262]. The calculation is subject to various criticisms, but the result must be assumed to be correct within an order of magnitude.

This means that phenomenally high reaction rates could be obtained with slurries of fine particles if the catalyst were so active that mass transfer of hydrogen to the particles were rate controlling. It would be difficult to approach hydrogen saturation in the liquid at such potentially high reaction rates, bu not impossible. This may be illustrated by taking Calderbank's representative values for the behavior of gas-liquid suspensions on a sieve plate at high gas flow rates ($d_B = 4.5$ mm, $H = 0.60$, bubble-liquid interfacial area $= 8\ cm^2/cm^3$ of expanded suspension). The value of k_L may be estimated from Equation 2.47.

$$N_{Sc} = \frac{0.0007}{(0.51)(5 \times 10^{-4})} = 2.74,$$

$$k_L = \frac{(0.42)}{(2.74)^{1/2}} \left[\frac{(3.0 - 0.51)(0.0007)(981)}{(0.51)^2}\right]^{1/3},$$

$$k_L = 0.42 \text{ cm/sec}.$$

The ratio of the maximum possible rate of mass transfer from bubble interface to bulk liquid to the maximum possible rate of mass transfer from bulk liquid to catalyst surface is $k_L A_B / k_c A_p$. (If gas-liquid mass transfer is controlling, the driving force is $(c_e - 0)$, and this is likewise true if the liquid is in equilibrium with the gas phase and mass transfer from liquid to catalyst particles is controlling.)

Expressing the particle surface area per unit volume of expanded slurry:

$$\frac{k_L A_B}{k_c A_p} = \frac{(0.42)(8)}{(2)(16 \times 0.4)} = 0.26.$$

It is apparent that the rate of mass transfer from bubble interface to liquid could be made comparable to the maximum obtainable from liquid to catalyst surface.

The high hydrogenation rate calculated in the preceding example checks roughly with an extrapolation of the data of Polejes, Hougen, and co-workers [254], for the hydrogenation of α-methylstyrene. At 15.5°C and 11.2 atm, with 0.89 g catalyst per liter, they measured a rate of 0.00149 g-mol hydrogen per liter per second. At 15.5°C, D_{1m} was about $5.2 \times 10^{-5}\ cm^2/sec$, and would be expected to be 5×10^{-4} or even higher at 316°C. The estimated rate for the conditions of the numerical example would be

$$\text{Rate} = 0.00149 \times \frac{5 \times 10^{-4}}{5.2 \times 10^{-5}} \times \frac{54.5}{11.2} \times \frac{8}{0.89}$$

$$= 0.63 \text{ g-mol } H_2/\text{sec} \cdot \text{l}.$$

Making no allowance for the different solvent, K at 316°C would be expected to be roughly one-sixth to one-eighth the value at 15.5°C (the solubility of hydrogen increases with temperature), indicating a rate of about 4 g-mol/sec · l. This is a rough check of the validity of the calculated value of 10.6 g-mol/sec · l.

Example 2.6 Application of Criteria for Absence of Diffusion Effects in a Slurry Reactor

Ma [297] studied the hydrogenation of α-methylstyrene to cumene at 70–100°C in an agitated reactor using a suspension of powdered catalyst consisting of 0.5 per cent Pd on alumina. The catalyst loading m was 4×10^{-3} g/cm³ slurry. It is believed that intrinsic kinetics were observed. The highest rate, observed at 100°C with pure α-methylstyrene was about 4.3×10^{-8} g-mol/cm³ slurry · sec.

The catalyst particles had an average diameter d_p of 0.005 cm and a density ρ_p of 1.58 g/cm³. At 100°C the density of the α-methylstyrene is 0.84. The viscosity was 0.47 cP. The bulk diffusivity of hydrogen in α-methylstyrene at 100°C is about 3.5×10^{-4} cm²/sec. The solubility of hydrogen in α-methylstyrene at 100°C is estimated to be about 3.6×10^{-6} g-mol/cm³.

Equation 2.44 gives the Stokes' law settling velocity:

$$U_{\text{term}} = \frac{g d_p^2 \Delta\rho}{18\mu}$$

$$= \frac{(981)(0.005)^2(1.58 - 0.84)}{(18)(0.0047)}$$

$$= 0.22 \text{ cm/sec.}$$

The corresponding Peclet number is

$$N_{\text{Pe}}^* = \frac{d_p U}{D_{1m}} = \frac{(0.005)(0.22)}{3.5 \times 10^{-4}} = 3.1.$$

From Figure 2.4, $k_c\, d_p/D_{1m}$ is only slightly greater than 2. Taking k_c as approximately double k_c^*, $k_c = (4)(3.5 \times 10^{-4})/0.005 = 0.28$ cm/sec. The rate of mass transfer of hydrogen from liquid to the pellet surface is

$$N = k_c \frac{6m}{d_p \rho_p} (c_H - c_{H,s}) \text{ g-mol/sec} \cdot \text{cm}^3.$$

If the rate were completely controlled by mass transfer, $c_{H,s}$ would approach zero and the observed rate would be

$$N = \frac{(0.28)(6)(4 \times 10^{-3})}{(0.005)(1.58)} (3.6 \times 10^{-6})$$

$$= 3.1 \times 10^{-6} \text{ g-mol/sec} \cdot \text{cm}^3.$$

This is about 70 times the fastest rate observed, so the hydrogen concentration gradient between bulk liquid and catalyst outside surface must have been negligible even if some of the numbers used in the calculation are somewhat in error.

The effectiveness factor for the catalyst powder at 100°C can be estimated by the methods of Chapter 3. The modulus Φ_s may be estimated as follows: The bulk diffusivity of hydrogen in α-methylstyrene in the porous catalyst is reduced by a factor of 8 to allow for porosity and tortuosity of the catalyst.

$$\Phi_s \equiv \frac{R^2}{D_{\text{eff}}}\left(-\frac{1}{V_c}\frac{dn}{dt}\right)\left(\frac{1}{c_s}\right) = \frac{(0.0025)^2(4.3 \times 10^{-8})}{(3.5 \times 10^{-4}/8) \times 3.6 \times 10^{-6}}$$

$$\approx 10^{-3}.$$

Since Φ_s is so small, the effectiveness factor for mass transfer inside the catalyst particle is essentially unity.

3 Diffusion and Reaction in Porous Catalysts I. Simple Treatment

3.1 Introduction

When reaction occurs simultaneously with mass transfer within a porous structure, a concentration gradient is established and interior surfaces are exposed to lower reactant concentrations than surfaces near the exterior. The average reaction rate throughout a catalyst particle under isothermal conditions will almost always be less than it would be if there were no mass transfer limitations. As will be shown, the apparent activation energy, the selectivity, and other important observed characteristics of a reaction are also dependent upon the magnitude of these concentration gradients.

The effects of mass transfer within a porous structure on observed reaction characteristics were apparently first analyzed quantitatively by Thiele [333] in the United States, Damköhler [85] in Germany, and Zeldovitch in Russia [397], all three working independently but reporting their results in the period 1937 to 1939. Thiele [332] has recently published an engaging historical account of how the concept of these effects gradually evolved. The early work was further developed by Wheeler [382, 383], by Weisz and co-workers [369, 372, 377], and by Wicke and co-workers [384–387]; it has since been extended by many others. The important result of these analyses is the quantitative description of the factors that determine the effectiveness of a porous catalyst. The *effectiveness factor*, here *symbolized by* η, is defined as the ratio of the actual reaction rate to that which would occur if all of the surface throughout the inside of the catalyst particle were exposed to reactant of the same concentration and temperature as that existing at the outside surface of the particle. The general theoretical approach is to develop the mathematical equations for simultaneous mass transfer and chemical reaction as the reactants and product diffuse into and out of the porous catalyst.

The mathematical analysis of one of the simplest systems will be considered in some detail to illustrate the approach. The effects of treating more complex situations will be considered in Chapter 4. Catalyst poisoning, selectivity effects caused by diffusion, and gasification of coke deposits from porous catalysts will be taken up in Chapter 5.

Consider a spherical catalyst pellet of radius R and focus attention on a spherical shell of thickness dr and radius r, as shown in Figure 3.1.

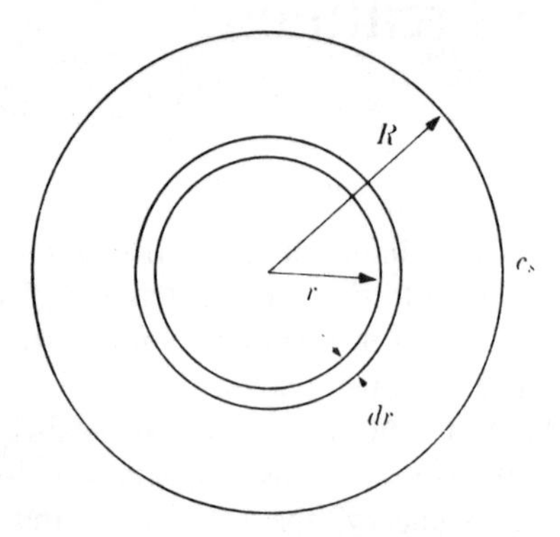

Figure 3.1. Spherical model for simultaneous diffusion and reaction.

We make the following assumptions for the simple treatment: (1) the porous structure is spherical in shape; (2) the porous structure is isothermal; (3) diffusion of a species within the porous structure can be represented by Fick's first law and an over-all invariant effective diffusion coefficient, i.e., flux $= -D_{\text{eff}}(dc/dr)$; (4) reaction involves a single reactant A, is irreversible, and the intrinsic rate of reaction may be represented by an integer power function of the concentration of A, i.e., as $k_s' c_A^m$; and (5) steady-state conditions. Reactants are transported to the differential annular shell by diffusion and consumed within it by reaction. *At steady state*, a mass balance on the differential shell becomes

(Rate of diffusion inward at $r = r + dr$)

— (Rate of diffusion inward at $r = r$)

= Rate of reaction in shell. (3.1)

Then Equation 3.1 becomes

$$4\pi(r + dr)^2 D_{\text{eff}}\left[\frac{dc}{dr} + \frac{d^2c}{dr^2}\,dr\right] - 4\pi r^2 D_{\text{eff}}\frac{dc}{dr} = (4\pi r^2\,dr)(S_v)k_s' c^m, \tag{3.2}$$

where $4\pi r^2$ is the inner superficial area of the spherical shell, D_{eff} is the effective over-all diffusion coefficient through the porous structure, dc/dr is the concentration gradient at the radius r, and S_v is the pore surface (cm^2/cm^3 of volume of the porous structure). The units of k_s' are (cm^3 of fluid)^m/(cm^2 of catalyst surface) (g-mol)$^{m-1}$(sec). Equation 3.2 becomes

$$D_{\text{eff}}\frac{d^2c}{dr^2}(4\pi r^2)\,dr + 8\pi r D_{\text{eff}}\frac{dc}{dr}\,dr = 4\pi r^2 S_v k_s' c^m\,dr, \tag{3.3}$$

or

$$\frac{d^2c}{dr^2} + \frac{2}{r}\frac{dc}{dr} = \frac{S_v k_s' c^m}{D_{\text{eff}}}. \tag{3.4}$$

This is to be solved for the boundary conditions of $c = c_s$ at $r = R$ (that is, the outside surface of the pellet) and $dc/dr = 0$ at $r = 0$ (that is, no concentration gradient will exist at the center of the sphere).

It is convenient to define a dimensionless quantity, termed the Thiele diffusion modulus ϕ_s,

$$\phi_s = R\sqrt{\frac{S_v k_s' c_s^{m-1}}{D_{\text{eff}}}} = R\sqrt{\frac{k_v' c_s^{m-1}}{D_{\text{eff}}}}, \tag{3.5}$$

where m is the order of reaction, and the subscript on ϕ_s is to remind us that we are dealing with spherical geometry.

Here k_v' $(= k_s' S_v)$ is the intrinsic reaction-rate constant per unit of gross volume of catalyst pellet. We consider now the case of a first-order reaction, where $m = 1$. The Thiele modulus then becomes independent of concentration, and Equation 3.4 may be written

$$\frac{d^2c}{dr^2} + \frac{2}{r}\frac{dc}{dr} = \frac{\phi_s^2}{R^2}c = b^2 c, \tag{3.6}$$

where $b = \phi_s/R$. To simplify, we set $c = v/r$, which converts Equation 3.6 to

$$\frac{d^2v}{dr^2} = b^2v. \tag{3.7}$$

The solution for Equation 3.7 is

$$v = cr = C_1e^{br} + C_2e^{-br}, \tag{3.8}$$

where C_1 and C_2 are constants of integration. Then,

$$c = \frac{1}{r}(C_1e^{br} + C_2e^{-br}). \tag{3.9}$$

Solving Equation 3.9 for the boundary condition of $r = 0$, $dc/dr = 0$ yields

$$C_1 = -C_2. \tag{3.10}$$

Thus,

$$c = \frac{C_1}{r}(e^{br} - e^{-br}) = \frac{2C_1}{r}\sinh br. \tag{3.11}$$

At $r = R$, $c = c_s$, whence

$$c_s = \frac{2C_1}{R}\sinh bR. \tag{3.12}$$

Combining Equations 3.11 and 3.12 gives

$$\frac{c}{c_s} = \frac{\sinh br}{(r/R)\sinh bR} = \frac{\sinh(\phi_s r/R)}{(r/R)\sinh \phi_s} \quad \text{(sphere, first order)}. \tag{3.13}$$

Equation 3.13 describes the concentration profile in the catalyst pellet and shows that this depends only upon the dimensionless parameter ϕ_s and the normalized distance r/R.

It is now desired to obtain the total reaction rate throughout the single catalyst pellet, corresponding to the concentration profile of Equation 3.13. This may be obtained by integrating the reaction rate in the spherical annulus over the entire sphere. A simpler solution proceeds, however, by recognizing that the over-all reaction rate equals the rate of mass transfer into the pellet:

$$\text{Rate} = 4\pi R^2 D_{\text{eff}}\left(-\frac{dc}{dr}\right)_{r=R}. \tag{3.14}$$

To determine the value of the concentration gradient at the outer surface of the pellet, we differentiate Equation 3.13 with respect to r:

$$\frac{dc}{dr} = \frac{c_s(\phi_s/R)\cosh[\phi_s(r/R)] - (c_s/r)\sinh[\phi_s(r/R)]}{(r/R)\sinh \phi_s}. \tag{3.15}$$

At the outer surface, this becomes

$$\begin{aligned}\left(-\frac{dc}{dr}\right)_{r=R} &= \frac{c_s[\phi_s \cosh \phi_s - \sinh \phi_s]}{R \sinh \phi_s} \\ &= \frac{c_s}{R}\left[\frac{\phi_s}{\tanh \phi_s} - 1\right] \\ &= \frac{\phi_s c_s}{R}\left[\frac{1}{\tanh \phi_s} - \frac{1}{\phi_s}\right].\end{aligned} \tag{3.16}$$

Substituting Equation 3.16 into Equation 3.14 gives

$$\text{Rate} = 4\phi_s \pi R D_{\text{eff}}\, c_s\left[\frac{1}{\tanh \phi_s} - \frac{1}{\phi_s}\right]\text{g-mol/sec.} \tag{3.17}$$

This is the rate of reaction throughout one spherical porous pellet corresponding to the concentration profile described by Equation 3.13.

If the internal surface of the porous pellet were all exposed to reactant of concentration c_s, the rate would be

$$\text{Rate} = \tfrac{4}{3}\pi R^3 k_v c_s \text{ g-mol/sec.} \tag{3.18}$$

The effectiveness factor η of the catalyst pellet is the ratio of the two rates given by Equations 3.17 and 3.18:

$$\eta = \frac{3}{\phi_s}\left[\frac{1}{\tanh \phi_s} - \frac{1}{\phi_s}\right] \quad \text{(sphere, first order)}, \tag{3.19}$$

where ϕ_s is defined by Equation 3.5. The function represented by Equation 3.19 is for *first-order irreversible reaction in a sphere.* The hyperbolic tangent approaches unity as ϕ_s is increased (it is 0.99 when $\phi_s = 2.65$) and η approaches $3/\phi_s$ at large values of ϕ_s.

Analytical solutions relating η to ϕ may be obtained in similar fashion for other simple orders of reaction and certain other geometries such as flat plates and cylinders. Figure 3.2 shows the functions for zero-, first-, and second-order reaction in a sphere and first-order reaction in a flat plate.

Equation 3.19 has important practical implications. The effectiveness

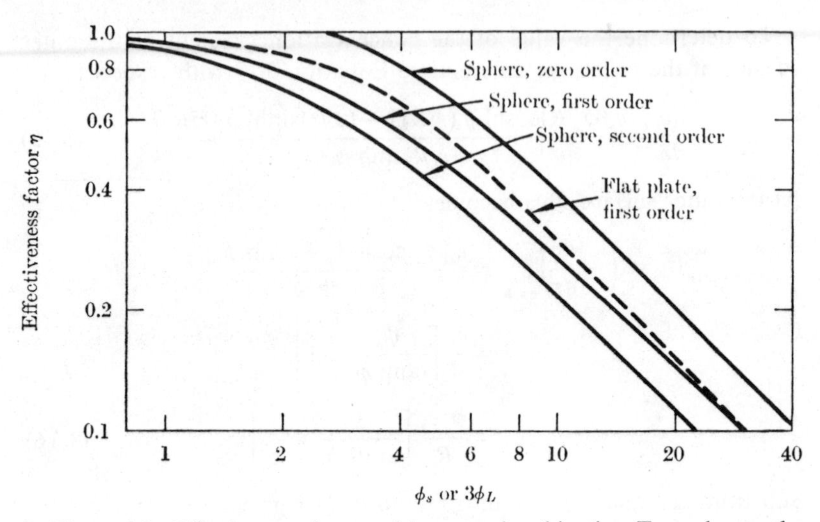

Figure 3.2. Effectiveness factor with power-law kinetics. For spheres, the abscissa is ϕ_s (Equation 3.5). For the flat plate, the abscissa is $3\phi_L$ (Equation 3.22).

of the internal surface of the porous catalyst approaches unity as an asymptote as the pellet radius or rate constant k_v is made small, or as the diffusion coefficient is made large. Conversely, the effectiveness becomes small for large particles, large rate constant, or small values of D_{eff}. The most effective use of a catalyst is attained by employing the smallest practical pellet size. Very active catalysts (k_v large) tend to have low effectiveness factors, while inactive catalysts tend to have high effectiveness factors. At large values of ϕ, the reaction rate per unit volume of catalyst increases only as the square root of k_v, as will be shown. For a given intrinsic kinetic expression and under isothermal conditions, the extent to which diffusion effects within the pellet are significant is determined solely by the value of the Thiele modulus and not by the individual values of R, k_v, or D_{eff}.

Zero-order reactions are unique in that reaction rate is independent of reactant concentration. Hence, $\eta = 1$ whenever the reactant concentration is finite at $r = 0$. Whenever $\eta < 1$, the reactant concentration becomes zero at some critical value of r, i.e., for all smaller values of r the reactant is completely exhausted. The concentration profile is given by [363]

$$c/c_s = 1 - (\phi_s^2/6)[1 - (r/R)^2] \quad \text{(sphere, zero order).} \qquad (3.20)$$

If $\phi_s < \sqrt{6}$, then $\eta = 1.0$ for zero-order reaction in spheres.

3.2 Geometries Other than a Sphere

The derivation in Section 3.1 is for a sphere, but the simplest geometry for mathematical analysis is a flat plate of porous catalyst in contact with reactant on one side but sealed on the other side and on the edges, frequently termed an *infinite* or *semi-infinite flat plate*. The differential equation to be solved is

$$D_{\text{eff}} \frac{d^2c}{dx^2} = c^m k_v', \tag{3.21}$$

where x is the distance from the sealed side. The Thiele modulus in this case is defined as

$$\phi_L = L\sqrt{k_v' c_s^{m-1}/D_{\text{eff}}}, \tag{3.22}$$

where L is the plate thickness. For a first-order irreversible reaction, the concentration term again disappears from the modulus and the solution is [333, 382, 383]

$$\eta = \tanh \phi_L/\phi_L \quad \text{(flat plate, first order).} \tag{3.23}$$

This is graphed in Figure 3.2.

The ratio of the intraparticle concentration to that at the outside surface is given by [333, 363]

$$c/c_s = \cosh(\phi_L x/L)/\cosh \phi_L \quad \text{(flat plate, first order).} \tag{3.24}$$

The gradient, in normalized form is

$$\frac{d(c/c_s)}{d(x/L)} = \phi_L\left[\frac{\sinh(\phi_L x/L)}{\cosh \phi_L}\right] = \phi_L[(c/c_s)^2 - (c_i/c_s)^2]^{1/2}, \tag{3.25}$$

where c_i is the concentration at the center.

The equivalent expressions for spheres are given by Equations 3.13 and 3.15. Equations 3.23, 3.24, and 3.25 also represent loose, nonporous catalyst particles spread in a flat boat with reactant sweeping across the top surface. Here D_{eff} would be the effective diffusivity of the bed of particles. For a flat plate of catalyst contacted on both sides, L is one-half the thickness.

In some treatments of diffusion and reaction within porous catalysts [333, 382, 383], an idealized pore structure comprising an assemblage of open cylinders of uniform and identical radius is assumed. Diffusion and reaction down a single cylindrical pore may then be treated mathematically. Here, Equations 3.21, 3.22, and 3.23 again apply, where L

is the length of the cylinder or pore, D_{eff} is the diffusivity coefficient for diffusion through the open cylinder, and k_v' is replaced by $2k_s'/r_e$, where k_s' is the intrinsic reaction-rate constant per unit surface area of catalyst, and $2/r_e$ is the ratio of surface to volume of the pore. In ordinary porous systems the diameter of a capillary of equivalent size to a pore will be so small that concentration in the radial direction of the capillary will be essentially uniform, but this may not be true in tubes where the radius is much larger. For systems of this type, an analysis of the effect of radial diffusion on the effectiveness factor has been published by Bischoff [36].

For more complex geometries, but simple power-law kinetics, the mathematics becomes more involved. Expressions have been published for a porous rod of infinite length (or with sealed ends) by Aris [11] and by Walker, Rusinko, and Austin [358] and for a porous rod with open ends [11]. The mathematics for the case of gases contacting the inside of a porous annulus have been developed by Wechsler [361], who also gives a mathematical analysis for flow of reactant into an open cylinder, opposed by a flow of inert fluid from the opposite direction. A similar theoretical treatment was published by Hawtin and Murdoch [137], who studied the air-oxidation of tubes of graphite in which air was forced through the porous wall. The pores were relatively large and the mass transfer was treated as the sum of the effects of bulk diffusion, bulk flow due to pressure drop, and the volumetric change on reaction.

The function $\sinh x$ is defined as $\frac{1}{2}(e^x - e^{-x})$ and $\cosh x$ as $\frac{1}{2}(e^x + e^{-x})$. Thus, for large and positive values of x, $\cosh x \approx \sinh x \approx \frac{1}{2}e^x$. For $x = 3$, $\cosh x \approx \sinh x \approx 10$, and for $x = 5$, $\cosh x \approx \sinh x \approx 74$. Examination of Equations 3.13 and 3.24 shows the marked concentration gradients that develop at values of ϕ_s or ϕ_L corresponding to effectiveness factors much less than unity. This is also brought out by Figure 3.3 (see discussion below). Equations 3.19 and 3.23 show that at sufficiently high values of ϕ_s or ϕ_L, η approaches a value of $3/\phi_s$ or $1/\phi_L$, respectively. This is likewise true for any order reaction, although the value of η at which these simple asymptotic relations become valid varies substantially with geometry and the nature of the kinetics (Sections 4.2 and 4.3). The awkwardness of dealing with a different definition of the Thiele modulus for each different geometry can be circumvented by defining the length parameter L in the modulus as the ratio of volume of the shape to the surface area through which reactants diffuse into the volume. For a sphere, the ratio of volume to surface is $\frac{1}{3}R$, so $\phi_L = \frac{1}{3}\phi_s$. For a specified kinetic expression, the functions of η versus ϕ_L for spheres and flat plates approach identity at high values of ϕ_L if the spherical pellet is characterized by ϕ_L rather than ϕ_s. This is true for all other geometries. (Physically it is evident that at low values of η only a very thin outside layer of any shape is involved, and this behaves essentially as a flat plate would. Under these conditions the reactant concentration at the pellet center is so low as to have no effect on the concentration gradient at the outside pellet surface.) The functions likewise approach identity at low values of ϕ_L, and the greatest spread in values of η for the two geometrical extremes of spheres and flat plates occurs at values of ϕ_L of about 1. Here η for a flat plate (the highest)

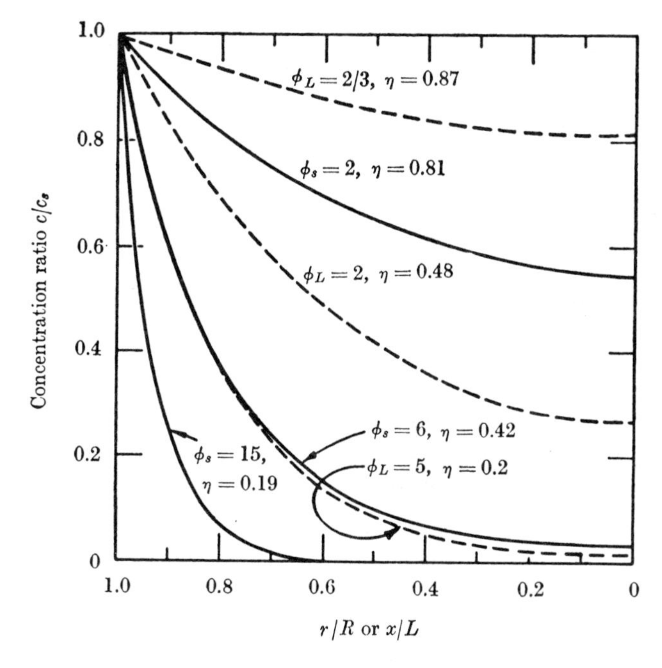

Figure 3.3. Concentration distribution in a porous spherical catalyst pellet and a flat plate. First-order reaction.

exceeds η for a sphere (the lowest) by about 0.09 for a first-order reaction as shown in Figure 3.2. The function of η versus ϕ_L is affected by the nature of the intrinsic kinetics as well as the geometry (Sections 4.2 and 4.3) but for most purposes the function as derived for one geometry can be used for any other with little error. Figure 3.3 displays the concentration distribution in flat plates and spheres as a function of the normalized distance coordinate, r/R or x/L, for three sets of pairs of values of ϕ_L and ϕ_s that are equivalent by the above approach (i.e., $\phi_L = \frac{1}{3}\phi_s$). For each set of equivalent ϕ_L and ϕ_s values the concentration profile in the sphere is substantially below that for the flat plate, but this is offset by the fact that a much greater fraction of the volume of a sphere is near the outside. Hence η is not markedly different for the two geometries. In a sphere 48.8 per cent of the volume is within the outside 20 per cent of the radius.

In the literature dealing with simple power-law kinetic relationships, ϕ_L as used here is symbolized as Λ by Aris, as ψ by Carberry, and as h by Wheeler. Tinkler and Metzner use $\sqrt{\alpha}$ to represent either ϕ_s or ϕ_L; ϕ_s is used in this book only for spherical geometry. It is symbolized as ϕ by Weisz and equals $3h$ as used by Wheeler. When dealing with more complex kinetics, some authors retain the same symbol used in earlier papers, but define it differently. Some authors use the symbols ϕ and Φ interchangeably, although in this book they have quite different meanings.

3.3 Some Characteristics of Diffusion-Limited Reactions

If ϕ_s is substantially less than about 1 ($\phi_L < \frac{1}{3}$), diffusion effects are relatively insignificant relative to kinetic effects, for first-, second-, or zero-order reactions, (but see Section 4.2 and Example 4.3). In this range the reactant concentration throughout the porous mass is essentially the same as that at the outside surface, and η is near unity. At values of η less than about 0.1, where η approaches $1/\phi_L$, substitution of the definition of the Thiele modulus into the reaction-rate expression yields

$$-\frac{1}{V_c}\frac{dn}{dt} = k_v' c_s^m \eta = \frac{c_s^{(m+1)/2}}{L}\sqrt{D_{\text{eff}} k_v'} \; \frac{\text{moles}}{\text{sec}\cdot\text{cm}^3 \text{ pellet volume}} \quad (\text{for } \eta < \sim 0.1). \tag{3.26}$$

Expressing the effect of temperature by the simple Arrhenius relationships, $k_v' = Ae^{-E/RT}$ and $D = Be^{-E_D/RT}$ the temperature-sensitive term $\sqrt{D_{\text{eff}} k_v'}$ becomes proportional to $e^{-(E+E_D)/2RT}$. It is seen that a catalyst operating at a low effectiveness factor will show an apparent activation energy on an Arrhenius plot that is the arithmetic mean of the activation energies for the diffusion process and for the chemical reaction.

$$E_{\text{obs}} = \tfrac{1}{2}(E + E_D). \tag{3.27}$$

In most gas-phase reactions, the activation energy for D_{eff} will be small relative to that for k_v', so the apparent activation energy will be about one-half that for the intrinsic reaction rate. The rate per unit gross volume of catalyst will be inversely proportional to the particle radius. An equivalent statement is that the rate per unit volume of catalyst pellets is directly proportional to the sum of the outside superficial areas of the catalyst particles. If the diffusion is in the Knudsen range, D_{eff} is independent of concentration and Equation 3.26 shows that an intrinsic second-order reaction will have an apparent kinetic order of 3/2 in this range. If ordinary bulk diffusion predominates in the catalyst pores, D_{eff} is inversely proportional to total pressure. If a single reactant is the only species present, with bulk diffusion in the pores, the measured rate is proportional to $p_s^{m/2}$. That is, an intrinsic reaction first order with respect to pressure would then show an apparent order of one-half and an intrinsic second-order reaction would show an apparent order of one.

Note that, even at low effectiveness factors, diffusion does not "control" the process since diffusion and reaction are not in series with one another, as is the case with mass transfer to the outside of a catalyst particle.

When pore diffusion is an insignificant resistance, a graph of the log of the reaction-rate constant versus the reciprocal of the absolute temperature is a straight line if k' can be expressed by a simple Arrhenius relationship. Its slope is equal to $-E/\mathbf{R}$. As the temperature is increased, however, η will gradually drop and the slope will decrease to the limiting value of about $-E/2\mathbf{R}$, as shown above. Weisz and Prater [377] have pointed out that the temperature region over which the apparent activation energy will vary may be as much as 200°C or so.

In this transition range the true activation energy is related to the observed value, E_{obs} by a relationship derived by Gupta and Douglas [129]:

$$E = \frac{2E_{obs} + E_D \psi}{2 + \psi}, \tag{3.28}$$

where $\psi = d(\ln \eta)/d(\ln \phi)$. ψ approaches zero at high values of η and -1 at low values, where Equation 3.28 reduces to Equation 3.27.

Topchieva, Antipina, and Shien [337] have reported a study of the effect of varying porosity, total surface, pore radius, etc., of 88 per cent silica–12 per cent alumina catalysts of varying activities on the kinetic parameters observed with the cracking of cumene. In agreement with theory, Arrhenius plots for the more active catalysts showed a pronounced break in the slope, indicating the transition to significant diffusion resistance. As the average pore was increased, the temperature at which the transition occurred also increased, again in agreement with theory, since D_{eff} will increase with pore size unless diffusion is completely by the bulk mode. The same type of break in an Arrhenius plot was reported by Vlasenko, Rusov, and Yuzefovich [350] for the reaction of carbon dioxide and hydrogen on a nickel catalyst. However, as discussed in Section 1.1, a marked drop in the apparent activation energy does not of itself prove that a diffusion limitation has developed; it may be caused solely by a shift in chemical mechanism.

If one is concerned with the reaction of two reactants A and B instead of a single reactant, the reaction may be highly diffusion limited with respect to one reactant but not to the second, as may readily occur with respect to hydrogen in the catalytic hydrogenation of a liquid [297]. Assuming the intrinsic kinetics to be first order with respect to both reactants, and the reaction to be highly diffusion limited with respect to A but not diffusion limited with respect to B, the Thiele modulus with respect to A is

$$\phi_{L,\mathrm{A}} = L\sqrt{\frac{k_v c_{\mathrm{B},s}}{D_{\mathrm{eff},\mathrm{A}}}}. \tag{3.29}$$

For $\eta_{\mathrm{A}} < \sim 0.1$, $\eta_{\mathrm{A}} \to 1/\phi_L$, and substituting into the reaction-rate expression,

$$-\frac{1}{V_c}\frac{dn}{dt} = k_v c_{\mathrm{A},s} c_{\mathrm{B},s}\eta = \frac{c_{\mathrm{A},s} c_{\mathrm{B},s}^{1/2}}{L}\sqrt{D_{\mathrm{eff},\mathrm{A}} \cdot k_v} \qquad (\text{for } \eta_{\mathrm{A}} < \sim 0.1). \tag{3.30}$$

Note that even though A is the diffusion-limited species, the apparent order of reaction becomes one-half *with respect to B* but remains first order with respect to A.

Equations 3.13 and 3.24 and Figure 3.3 show that when a catalyst is operating at a low effectiveness factor (high values of ϕ_s or ϕ_L) the reactant concentration drops rapidly with distance of penetration into the interior. Hence, the catalyst in the interior contributes relatively little to the over-all rate of reaction. If the active ingredient of the catalyst is expensive, as is the case with substances such as platinum or palladium, it may be deposited in a thin layer on the outside portion of the pellet to maximize its usefulness. It will be shown later that with complex reactions selectivity is almost always diminished if the catalyst effectiveness factor is much less than unity. In some cases a catalyst may be so active that to obtain an effectiveness factor near unity, and hence high selectivity, the requisite particle size is so small as to be impractical. In such cases the problem may be resolved by using a thin layer of active catalyst on an inert or nonporous carrier of suitable dimensions.

3.4 Determination of Effectiveness Factor

3.4.1 *Experimental*

The effectiveness factor can be determined most readily by measuring reaction rates on several catalyst particle sizes under otherwise identical conditions. The effectiveness factor approaches unity when no increase in rate per unit quantity of catalyst occurs on subdivision. If data are available on finely divided catalyst having an effectiveness factor approaching unity, then the ratio of the rate per unit quantity of catalyst for the larger size to that for the finer is directly equal to the effectiveness factor of the larger size.

If experimental rate data are available on only two particle sizes, the effectiveness factor may still be determined without independent knowledge of either k_v' or D_{eff}. Let the two particle radii be R_1 and R_2. The ratio of the rates per unit volume of catalyst on the two particle sizes as determined experimentally is η_1/η_2. The ratio R_1/R_2 is ϕ_1/ϕ_2. The simultaneous equations expressing η_1 as a function of ϕ_1 and η_2 as a function of ϕ_2 can then be solved to obtain η_1 and η_2. A graphical solution [377] is based on the fact that on a logarithmic plot of η versus ϕ, such as Figure 3.2, the ratio η_1/η_2 forms a line of fixed length on the ordinate and the ratio ϕ_1/ϕ_2 another fixed length on the abscissa. These two lengths are used to form a triangle that can be fitted to the curve, yielding the individual values of η_1, η_2, ϕ_1, and ϕ_2. For accuracy, however, the equivalent analytical expression, such as Equation 3.19 or Equation 3.23, should be used instead. These methods do not apply if both catalyst particle sizes are so large than η is small and the log η versus log ϕ curve becomes a straight line. In this region, the rate per unit catalyst volume is inversely proportional to catalyst particle diameter for all particle sizes.

3.4.2 *Theoretical. The Modulus* Φ

Frequently one has data on the rate of reaction on a particular catalyst of one fixed size and it is desired to estimate the extent to which a diffusion effect may be present, as represented by the effectiveness factor. The modulus ϕ contains the intrinsic kinetic constant k_v' which is usually unknown. An approach that greatly simplifies analysis is to use relationships between η and a new dimensionless modulus Φ defined as

$$\Phi_s \equiv \frac{R^2}{D_{\text{eff}}}\left(-\frac{1}{V_c}\frac{dn}{dt}\right)\frac{1}{c_s} \quad \text{(spherical geometry)} \qquad (3.31)$$

or

$$\Phi_L \equiv \frac{L^2}{D_{\text{eff}}}\left(-\frac{1}{V_c}\frac{dn}{dt}\right)\frac{1}{c_s} \quad \text{(flat plate geometry).} \qquad (3.32)$$

In this approach, first suggested by Wagner [351] and used by Weisz and co-workers, we utilize a modulus in which all the terms may be either observed or calculated. The parenthetical expression is the *observed* rate of reaction per unit volume of catalyst pellets; the other terms are the reactant concentration at the outside surface of the catalyst pellets, the radius (or other dimension) of the catalyst, and the effective diffusivity in the porous catalyst. It is still necessary to know the correct form of the intrinsic kinetic expression, but this can be postponed to the last stage of the calculation and available generalized charts (see Sections 4.2 and 4.4) make it possible to relate one's confidence in the resulting calculated effectiveness factor to one's confidence in the validity of the kinetic expression.

For any *integer-power* rate equation of order m,

$$\Phi_L = \phi_L^2\eta \quad \text{or} \quad \Phi_s = \phi_s^2\eta \quad \text{(integer-power kinetics).} \qquad (3.33)$$

The relationship between Φ_L and ϕ_L or Φ_s and ϕ_s is more complex for more complicated kinetic expressions. Equation 3.33 can be shown by formulating the rate of disappearance of reactant and eliminating k_v' by use of the definition of the Thiele modulus (e.g., Equation 3.5).

$$-\frac{1}{V_c}\frac{dn}{dt} = k_v' c_s^m \eta = \frac{c_s D_{\text{eff}}}{R^2}\phi_s^2\eta. \qquad (3.34)$$

If the order of the reaction *is* known and is of an integer power, then when $\eta < \sim 0.1$, $\eta = 1/\phi_L = 1/\Phi_L$ (for spheres, $\eta = 3/\phi_s = 9/\Phi_s$). For higher values of η, η may be determined by successive approximations from Figure 3.2 or a graph of η versus $\phi^2\eta = \Phi$ may be prepared to give η directly. Such a graph for first-order reaction in a sphere is given by the curve marked $\beta = 0$ in Figures 3.4–3.7. (These figures are discussed in more detail in Chapter 4.). D_{eff} can be estimated by the methods of Chapter 1. In cases where the resistance to mass transfer from fluid

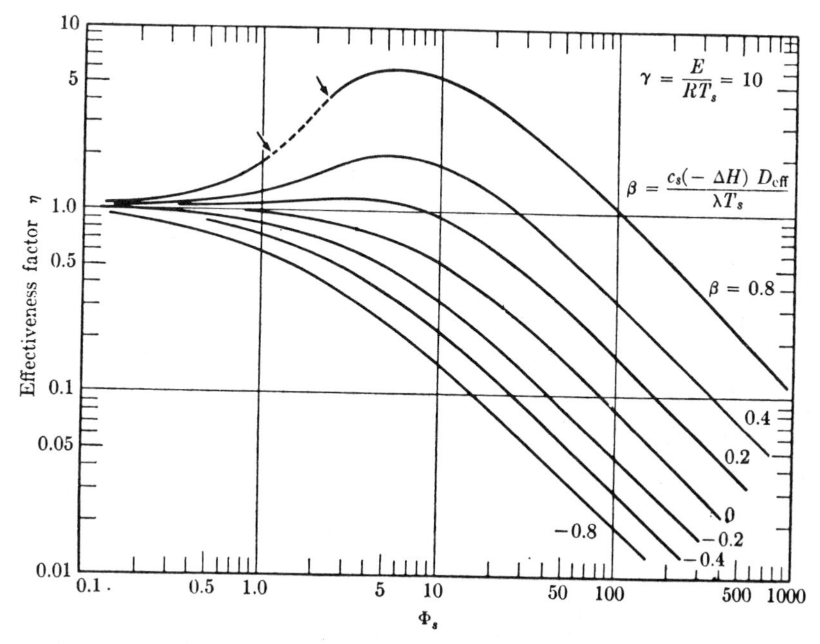

Figure 3.4. Effectiveness factor η as a function of Φ_s (Equation 3.31) for $\gamma = 10$. First-order reaction in sphere. Weisz and Hicks [376].

stream to outer pellet surface is negligible, c_s is replaced by c_0, the concentration of reactant in the ambient fluid.

Example 3.1 illustrates the method of calculating the effectiveness factor using the modulus Φ_s.

Example 3.1 Estimation of Effectiveness Factor η: Isothermal, First-Order Kinetics

Archibald, May, and Greensfelder [9] studied the rate of catalytic cracking of a West Texas gas oil at 550° and 630°C and 1 atm pressure by passing the vaporized feed through a packed bed containing a silica-alumina cracking catalyst of each of several sizes ranging from 8 to 14 mesh to 35 to 48 mesh. They report that at 630°C the apparent catalyst activity was directly proportional to the external surface for the three larger catalyst sizes studied. This implies that the catalyst is operating at a relatively low effectiveness factor. We can check this by suitable calculations based on their run on 8 to 14 mesh catalyst at 630°C. The average particle radius may be taken as 0.088 cm.

They report 50 per cent conversion at a liquid hourly space velocity (LHSV) of 60 cm³ of liquid per cm³ of reactor volume per hour. The liquid

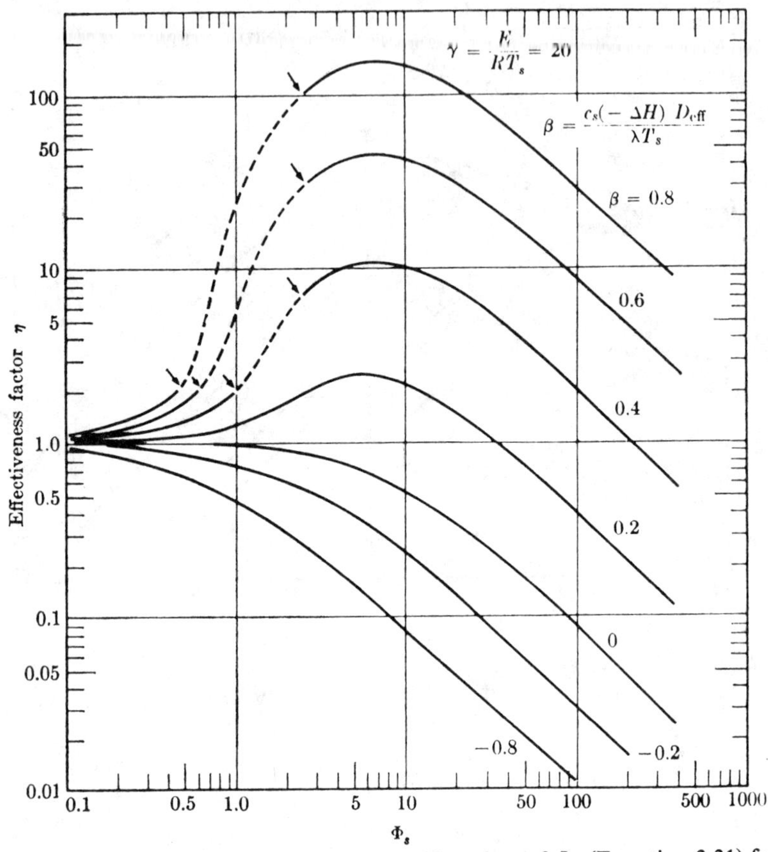

Figure 3.5. Effectiveness factor η as a function of Φ_s (Equation 3.31) for $\gamma = 20$. First-order reaction in sphere. Weisz and Hicks [376].

density is 0.869, and its average molecular weight is 255. The effective density of the packed bed was about 0.7 g catalyst/cm³ reactor volume. The molecular weight of the products was about 70.

Let us take pore-structure characteristics for a commercial silicia-alumina cracking catalyst to be those reported by Johnson, Kreger, and Erickson [159], viz., average pore radius $= 28$ Å, catalyst particle density $\rho_p = 0.95$, $\theta = 0.46$, $S_g = 338$ m²/g, τ is about 3.0 (see Table 1.11). From Equation 1.33,

$$D_{\text{eff}} = 19\,400 \frac{\theta^2}{\tau S_g \rho_p} \sqrt{\frac{T}{M}}$$

$$= \frac{(19\,400)(0.46)^2}{(3)(338 \times 10^4)(0.95)} \sqrt{\frac{903}{255}} = 8.0 \times 10^{-4} \text{ cm}^2/\text{sec}.$$

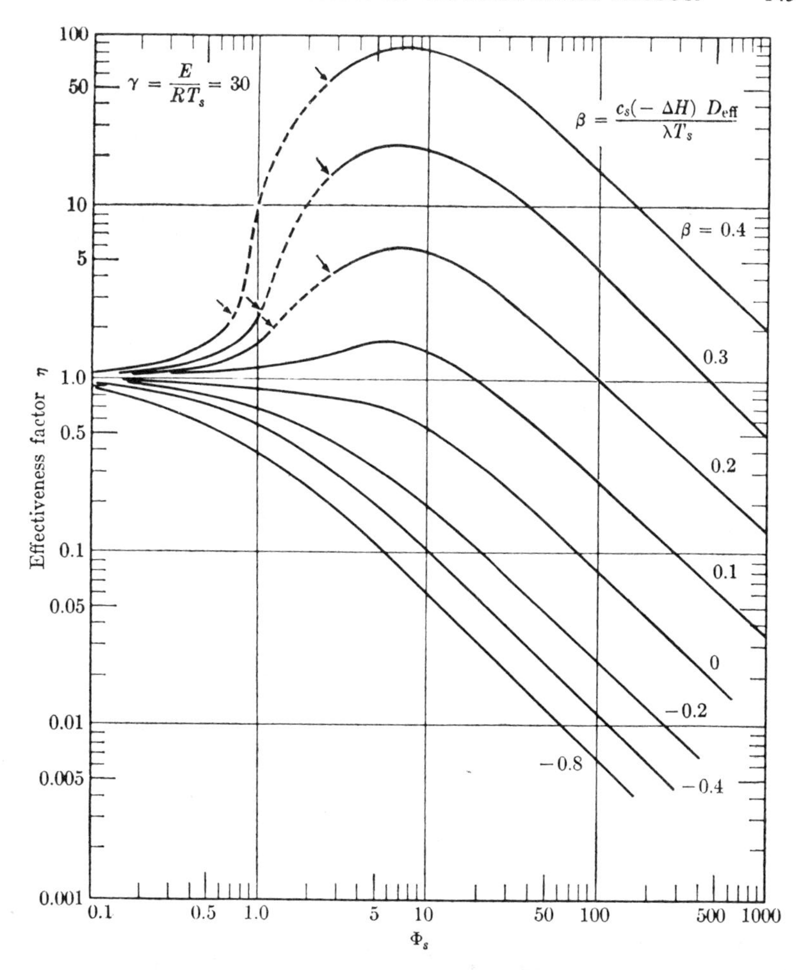

Figure 3.6. Effectiveness factor η as a function of Φ_s (Equation 3.31) for $\gamma = 30$. First-order reaction in sphere. Weisz and Hicks [376].

The rate of reaction is 50 per cent of the rate of feed

$$\left(-\frac{1}{V_c}\frac{dn}{dt}\right) = \frac{(60)(0.869)}{255}\left(\frac{1}{3600}\right)\left(\frac{1}{0.7}\right)(0.95)(0.5)$$

$$= 3.86 \times 10^{-5} \text{ g-mol/sec} \cdot \text{cm}^3 \text{ pellet volume.}$$

For the average concentration of reactant outside the pellets, an arithmetic average of inlet and exit concentrations is sufficiently precise for this example. This corresponds to conditions at 25 per cent conversion.

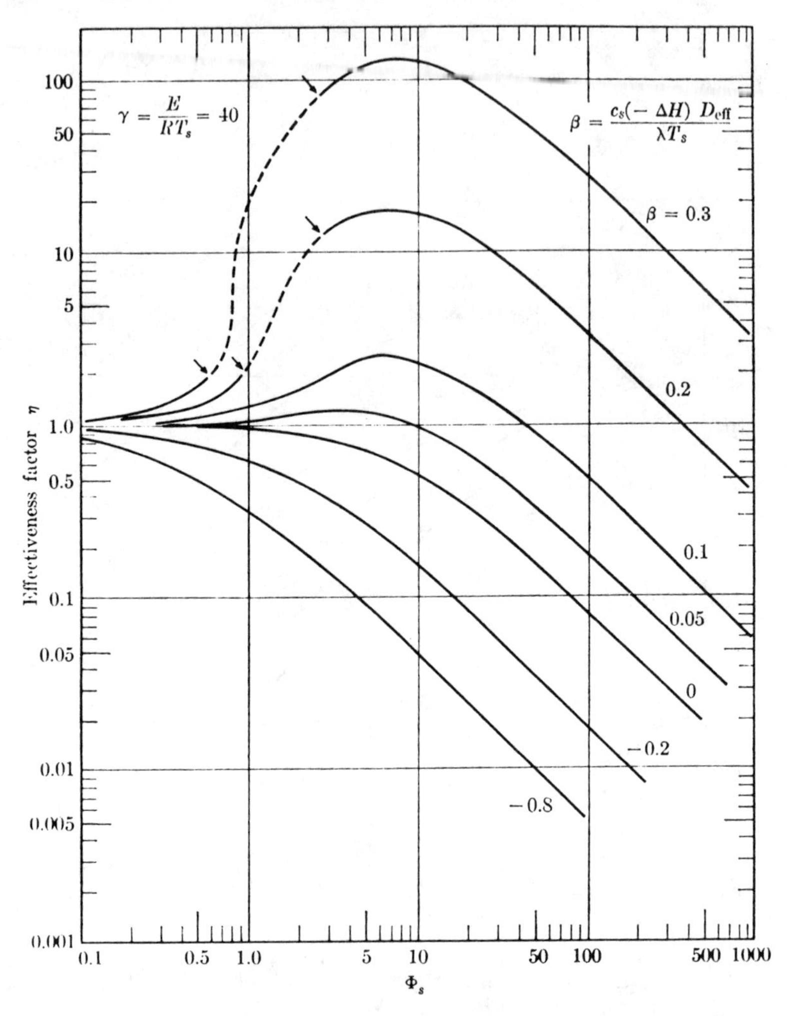

Figure 3.7. Effectiveness factor η as a function of Φ_s (Equation 3.31) for $\gamma = 40$. First-order reaction in sphere. Weisz and Hicks [376].

Each mole of gas oil produces $255/70 = 3.64$ moles of products. Per 100 moles entering, at 25 per cent conversion there remain 75 moles of gas oil and $(25)(3.64) = 91$ moles of products, for a total of 166 moles.

The average reactant concentration is therefore

$$\frac{1}{22\ 400}\left(\frac{273}{903}\right)\left(\frac{75}{166}\right) = 0.61 \times 10^{-5}\ \text{g-mol/cm}^3.$$

The average catalyst particle radius is 0.088 cm, whence, from Equation 3.31,

$$\Phi_s = \frac{(0.088)^2}{8 \times 10^{-4}} (3.86 \times 10^{-5}) \times \frac{1}{0.61 \times 10^{-5}} = 61.$$

For moderate degrees of reaction, the rate is approximately first order. From the curve for $\beta = 0$ from either Figure 3.4, 3.5, 3.6, or 3.7, the corresponding value of η is about 0.12.

3.4.3 *Integral Reactor*

The above analyses apply strictly to the case in which c_s is constant, i.e., to a differential reactor. For irreversible first-order kinetics the effectiveness factor for an integral reactor may be calculated by an analysis similar to that used in Section 2.3 for determining whether mass transfer to the external surface of catalyst pellets in a reactor is a significant limitation. Consider a differential section of a packed bed, having unit cross section and length dz in the direction of flow.

Then,

$$(1 - \varepsilon)\left(-\frac{1}{V_c}\frac{dn}{dt}\right) dz = -G_M\, dY, \tag{3.35}$$

where ε is the void fraction in the packed bed, and hence $1 - \varepsilon$ is the ratio of pellet volume to volume of reactor, z is the length in cm, and G_M is the molar fluid velocity based on bed superficial cross section. We substitute $k_v c_s \eta$ for $(-1/V_c\, dn/dt)$ (Equation 3.34) and the relationship PY/RT for c_s. For $\eta < \sim 0.1$, $\eta \approx 3/\phi_s$ and $k_v = 9D_{\text{eff}}/R^2\eta^2$.

Substituting in Equation 3.35 we obtain

$$\frac{(1 - \varepsilon)}{G_M}\left(\frac{P}{RT}\right)\frac{9}{R^2}\frac{D_{\text{eff}}}{\eta}\, dz = -\frac{dY}{Y}. \tag{3.36}$$

This neglects mass transfer resistance from flowing gas to pellet surface, so the mole fraction Y in the gas stream is taken equal to the mole fraction Y_s at the pellet surface.

If the terms in the left-hand side of the equation do not vary with length, integration of 3.36 provides an expression that enables the effectiveness factor to be calculated from readily obtained data on the bed as a whole.

$$\eta = \frac{9}{R^2}\frac{(1 - \varepsilon)}{G_M}\left(\frac{P}{RT}\right)\frac{D_{\text{eff}}\, z}{\ln(Y_1/Y_2)}. \tag{3.37}$$

Here Y_1 and Y_2 are the mole fractions of the reactant entering and leaving the reactor, respectively.

Alternatively, the effectiveness factor may be related to a nominal residence time $t' = (z\varepsilon/G_M)(P/RT)$. Equation 3.37 thus becomes

$$\eta = 9\frac{(1-\varepsilon)}{\varepsilon}\left(\frac{D_{\text{eff}}}{R^2}\right)\frac{t'}{\ln(Y_1/Y_2)}. \tag{3.38}$$

Here t' is the nominal residence time at reactor conditions, based on the void fraction in the bed external to the particles, assuming no change in linear velocity occurs with length of bed. A space velocity (S.V.) may be defined as the volumetric flow rate divided by the bed volume, which is $G_M RT/Pz$, whence,

$$\eta = \frac{9(1-\varepsilon)D_{\text{eff}}}{R^2 \ln(Y_1/Y_2)(\text{S.V.})}. \tag{3.39}$$

Note that Equations 3.36–3.39 are only valid for low values of η, e.g., $\eta < \sim 0.1$, and for first-order reactions, for which ϕ_s does not involve c_s, so η is constant through the bed. The integration is slightly more complicated in the case of a reaction of higher order, since η then varies with Y and must be eliminated to express z as a function of Y_1 and Y_2.

3.4.4 *Particle-Size Distribution*

If a distribution of catalyst particle sizes exists, Aris [11] showed that the mean effectiveness factor is given by

$$\eta = \eta_1 f_1 + \eta_2 f_2 + \cdots, \tag{3.40}$$

where f_1, f_2, etc., are the volume fractions of particles of sizes 1, 2, etc. The volumetric reaction rate $(-1/V_c)(dn/dt)$ varies with particle size if $\eta < 1$. If the effectiveness factor is sufficiently small over the entire particle-size distribution that η is inversely proportional to ϕ, and a log-normal distribution of particle sizes exists, then the observed volumetric reaction rate equals that occurring on a uniform particle size of radius R, where

$$R = \bar{R}\exp[\tfrac{1}{2}(-\log^2 \sigma)]. \tag{3.41}$$

Here $\bar{R}$ is the weight geometrical mean particle size and σ is the standard deviation of the log-normal distribution of particle sizes [114].

3.5 Effectiveness Factors in Ion Exchange Resins

Cation exchange resins, such as those containing sulfonic acid groups, may be used in the acid form to catalyze reactions that are accelerated by hydrogen ions. Such reactions include the inversion of sucrose, esterification reactions, and the hydrolysis of esters. The rate of such reactions can become limited by the rate of diffusion of reactants into the resin particles, even when the particles are of sizes of the order of 1 mm or less [38, 325]. Most of the reported studies used copolymer beads of styrene cross-linked with a small amount, e.g., 8 per cent, of divinyl benzene which was then sulfonated (Dowex-50, Amberlite IR-100 or IR-120). These materials are similar to inorganic zeolites ("molecular sieves") but their pore structure is more difficult to characterize since the gels swell in solution and the extent of swelling varies with degree of cross-linking of the polymer, the nature of the solvent, and other factors.

The effectiveness factors for such systems can be determined by methods analogous to those used in other porous catalysts. In an early study, Smith and Amundsen [325] measured the rate of hydrolysis of dilute solutions of ethyl formate at 25°C on Dowex-50 resin in five particle sizes ranging from 20 to 30 mesh to one smaller than 230 mesh. From the effect of particle size on reaction rate, they calculated $\sqrt{k/D_{eff}}$ to be 137.8 cm^{-1}, and D_{eff} to be 5.0×10^{-5} cm^2/min. They noted that this value is about one-tenth of the value of D_{1m}. If the average size of the 20 to 30 mesh particles were taken as 0.7 mm diameter, the effectiveness factor for this size becomes 0.62.

Bodamer and Kunin [38] studied the rate of inversion of 20 per cent sucrose in water at 50°C on Amberlite IR-120 (which contains about 8 per cent divinyl benzene) in particle sizes down to 0.24 mm. Even on the smallest particles, the rates were diffusion limited. If the effectiveness factor is expressed as the ratio of the rate on the 0.24 mm particles to that occurring homogeneously in the presence of 0.18 *N* hydrochloric acid, which has the same hydrogen ion concentration as the resin, η becomes 0.044, although this method of comparison may not be completely valid. O'Connell [230] also studied the inversion of sucrose in aqueous solution at 50°, 60°, and 70°C on four sizes of resin beads (0.04, 0.27, 0.55, and 0.77 mm diameter) of Dowex 50W × 8, which also contains about 8 per cent divinyl benzene. The data fitted well the theoretical relationship for diffusional limitations for a first-order irreversible reaction in a sphere and the effectiveness factors varied from essentially unity for the finest particles to about 0.1 for 0.77 mm spheres at 70°C. For 0.27 mm particles at 50°C, η was about 0.55, indicating that the Dowex 50W had a considerably less restricting pore structure than the similar polymeric resin used earlier by Bodamer and Kunin [38]. In O'Connell's study separate measurements of D_{eff} by transient adsorption of sucrose into the sodium form of the resin agreed well with values back-calculated from the reaction rate studies. D_{eff} was 27×10^{-8} cm^2/sec at 50°C, which is about one-tenth the value for bulk diffusion in water. The activation energy for diffusion was 5.5 kcal/g-mol and that on the finest beads, which represents intrinsic kinetics, was 25 kcal. The activation energy on the 0.077 cm beads was 18 kcal/mol, which is close to that which would be predicted from Equation 3.27 for a highly diffusion-limited system. In a study of sucrose inversion on Amberlite IR-120 at 50° to 75°C, Reed and Dranoff [263] reported an inverse relationship between rate and particle size and an activation energy of 16 kcal/mol on 16–20 mesh resin sizes, both of which indicate significant diffusion limitations. On Amberlite IR-100 resin, hydrolysis of low-molecular-weight esters apparently proceeds at relatively high effectiveness factors, but esterification of butanol and oleic acid, the latter a large molecule, involves a low effectiveness factor [38].

The pore structure of these gels collapses on drying, since the B.E.T. area of dried samples is less than 0.1 m^2/g [183]. By changing the degree of cross-linking, Bodamer and Kunin were able to change the gel structure and thus vary the observed rate constant by more than 200-fold for a given particle size. This approach has been pursued further by the preparation of so-called macroreticular ion exchange resins, which have a rigid macroporous structure superimposed on the gel structure [114, 183, 184]. These resins are prepared by copolymerizing polystyrene and divinyl benzene in the presence of a substance that is a good solvent for the monomer but a poor swelling agent for the polymer. Acid groups are then introduced by sulfonation. This macropore structure does not disappear on drying. The B.E.T. surface area of one such resin (Amberlyst-15) is 42 m^2/g and the pore diameter about 288 Å. Such a resin when wet has a macro-micro pore distribution analogous to those encountered in many conventional catalysts.

For at least some of these reacting systems the mechanism of mass transfer through the gel seems to be more similar to surface diffusion than to a bulk diffusion flux reduced by porosity and tortuosity of the gel. Thus, a study of isobutylene hydration on Dowex-50 under diffusion-limited conditions by Gupta and Douglas [128] showed that the effective diffusivity as back-calculated from reaction rate data *decreased* from 5.5×10^{-5} at 74°C to 1.7×10^{-5} at 89°C. This is reminiscent of the study of Bienert and Gelbin [29] on dehydration of isopropanol on γ-alumina in which the effective diffusivity back-calculated from reaction-rate studies reached a maximum value at a temperature intermediate between the extremes of 200° and 250°C studied. Both studies strongly suggest that the predominant mode of mass transfer was by a surface diffusion mechanism.

Diffusion limitations in ion exchange resin catalysts have also been studied by Saito *et al.* [289] and Fang [104], and are discussed in the books on ion exchange resins by Kunin [182] and Helfferich [139]. For many reactions in these resins the intrinsic kinetics may be closely represented as a simple first-order reversible reaction. Frisch [114] presents analytical solutions for this case. See also Section 4.4. If the diffusing species are charged, as in ion exchange processes, their diffusion under a concentration gradient produces an electric field which may in turn significantly alter the flux.

3.6 Validation of Theory. Isothermal Reaction

Table 3.1 shows effectiveness factors for a number of catalytic reactions occurring in a porous solid. With a few exceptions, the list includes only studies in which the effectiveness factor was determined by the straightforward experimental method of varying particle size, and of reactions or on catalysts of industrial interest. In some cases, effective diffusivities in the absence of reaction and other information on the catalyst structure were provided from which the mathematical theory for prediction could be tested. These are discussed below together with other studies not included in Table 3.1 which were aimed primarily at testing theory.

The principal uncertainties in predicting the effectiveness factor in practice seem to stem from one's degree of confidence in predicting the effective diffusivity under reaction conditions, one's knowledge of the nature of the intrinsic kinetics, and the degree of anisotropy which exists in the catalyst structure.

The effect on the effectiveness factor of the nature of the kinetic expression that governs the rate of reaction has now been worked out in considerable detail, assuming that the intrinsic activity of the catalyst is uniform through the structure. Anisotropies in the catalyst structure can be grouped into two kinds, those affecting the local value of the effective diffusivity, such as skin effects, fine cracks, the incorporation

Table 3.1 Some Effectiveness Factors Reported in the Literature

Catalyst	Reaction	Reaction Conditions	Catalyst Characteristics	Pellet Size (diam in cm)	Effectiveness Factor η	Author and Method
1. Silica-alumina pellets	Cracking of gas oil	500°C, 1 atm	$d = 20–100$ Å	0.32	→1.0	Wheeler [383], by effect of particle size
2. Silica-10% alumina pellets	Cracking of gas oil	480°C, 1 atm	Commercial bead	0.44	0.55	Johnson, Kreger, and Erickson [159], by subdivision of pellets
			Bead hydrogels prepared in various ways	0.44	0.60–0.79	
3. Silica-alumina pellets	Cracking of gas oil	630°C, 1 atm	Assumed $d = 28$ Å, calculated $D_{eff} = 8 \times 10^{-4}$ cm^2/sec	8–14 mesh	0.1	See Example 3.2. Data from Archibald, May, and Greensfelder [9]
4. Silica-alumina pellets	Cracking of cumene	420°C, 1 atm	$d = 40$ Å, measured $D_{eff} = 1.2 \times 10^{-3}$ cm^2/sec	0.46 0.35 0.01	0.15 0.2 →1.0	Prater, Weisz, and Lago, in Weisz and Prater [377], by variation of particle size
5. Silica-alumina TCC beads	Cracking of cumene	510°C, 1 atm	$S_g = 342$ m^2/g, $\theta = 0.41$	0.045 0.33 0.43 0.53	0.72 0.16 0.12 0.09	Corrigan *et al.* [76] by variation of particle size, η recalculated by Prater and Lago [257]
6. Silica-alumina TCC beads	Hydrogen transfer from decalin to butylene	342°C	$d = 48$ Å	0.5 40–60 mesh	0.12 ~0.93	Blue *et al.* [37], by effect of particle size. Analyzed by Wheeler [383]

7. Silica-alumina TCC white beads	Dehydration of ethanol, *n*-propanol or *n*-butanol	1 atm	$S_g = 350\ m^2/g$, $\rho_p = 1.15$, $\rho_t = 2.3$, $\theta = \sim 0.50$			Miller and Kirk [217] by variation of particle size
		260°C		0.5	0.15	
				0.23	0.24	
				0.04	0.98	
		370°C		0.5	0.10	η is independent of particular alcohol studied
				0.23	0.16	
				0.04	0.90	
8. Silica-alumina used in 4	Oxidation of coke deposited on used catalyst	O_2 conc'n = 3×10^{-6} mol/cm³, 1 atm	D_{eff} for $O_2 = 4 \times 10^{-3}$ cm²/sec	0.40	→1 at 460°C; $\eta < 1$ when $T \geq 490°C$	Weisz [372], effect on rate of variation of temp and particle size
				0.02 (150 mesh)	→1 at all temps studied (460–560°C)	See also Example 4.7
9. Chromia-alumina pellets	Dehydrogenation of cyclohexane	478°C, 1 atm	$D_{eff} = 8 \times 10^{-4}$ cm²/sec, computed from reaction data	0.62	0.48	Weisz and Swegler [379], by subdivision of pellets
				0.37	0.65	
10. Pt on alumina (% Pt unspecified)	Dehydrogenation of cyclohexane to benzene	340–490°C 200 psig mole ratio of H_2/cyclohexane = 4.0	$S_g = 240\ m^2/g$, $\theta = 0.59$, $\rho_p = 1.33$, $\rho_t = 3.35$, data best fitted by $\tau = 8$	0.3	$\eta < 1$ at $T > 370°C$, $\eta = 0.38$ at 430°C	Barnett, Weaver, and Gilkeson [16], by variation of particle size
11. Chromia (12%)-alumina pellets	Dehydrogenation of butane	530°C	$d = 220$ Å	0.32	0.70	Blue *et al.* [37], by particle size variation[a]

[a] A commercial chromia-alumina catalyst used for dehydrogenation of butane to butadiene in a cyclic industrial reactor will typically decrease in surface area from 55–60 m²/g when fresh to about 13 m²/g after use for a year. Corresponding average pore diameters are about 250 and 1100 Å. Thus, the effectiveness factor may increase substantially as the catalyst ages.

Table 3.1—*Continued*

Catalyst	Reaction	Reaction Conditions	Catalyst Characteristics	Pellet Size (diam in cm)	Effectiveness Factor η	Author and Method
12. Iron-alumina pellets (10.2% Al_2O_3, no K_2O)	Decomposition of NH_3	387–467°C, 1 atm	$d = 460$ Å	10–14 mesh to 35–40 mesh	→1.0	Love and Emmett [192], η independent of particle size
13. Iron-alumina pellets	Synthesis of NH_3	450–550°C. 100–600 atm	Bulk diffusion, $D_B = 0.674$ cm²/sec	0.5	$\eta > 0.8$	Calculated by Wheeler [38[illegible]], also η independent of particle size
14. Promoted iron catalyst (2.9% Al_2O_3 and 1.1% K_2O)	Synthesis of NH_3	450°C, 1 atm	$S_g = 13.1$ m²/g $r_e = 150$–180 Å	0.24–0.28 0.05–0.07	0.05 0.36	Bokhoven and van Raayen [39], by variation of particle size.
		325°C, 1 atm		0.24–0.28 0.05–0.07	0.78 →1	Also calc'd from D_{eff} measured at 1 atm, 20°C, with $N_2 - O_2$. See also Peters and Krabetz [247]
		500°C, 30 atm		0.24–0.28 0.05–0.07	0.40 0.89	
		400°C, 30 atm		0.24–0.28 0.05–0.07	0.95–1.0 →1	
15. Iron, with 2.5–3.5% promoter	Synthesis of NH_3	400–450°C, 1 atm		0.14–0.24	0.20	Data of Larson and Tour [187], η calculated by Bokhoven and van Raayen [39]
		425–450°C, 100 atm		0.24–0.33	0.80	
16. Industrial catalyst (?)	Synthesis of NH_3	450–480°C, 330–360 atm		0.12–0.23 1.0	0.89–0.98 0.15–0.40	Data of Nielsen [227], η calculated by Bokhoven and van Raayen [39]

17. Supported Ni catalyst	$CO_2 + H_2$	142°C 300°C	$S_g = 172\ m^2/g$	0.4	$\rightarrow 1.0$ ~ 0.1	Vlasenko, Rusov, and Yuzefovich [350]
18. 5 wt % Ni on γ-alumina pelleted from powder	$oH_2 \rightarrow pH_2$	−196°C, 100–400 psig	$S_g = 155\ m^2/g$ $\rho_p = 1.91$	0.4	0.43–0.45	Wakao, Selwood, and Smith [354], by variation of particle size
19. Porous nickel film deposit	Hydrogenation of ethylene	23°C	$d = 200$ Å $S_g = 16\ m^2/g$ $D_K = 1.77 \times 10^{-2}\ cm^2/sec$	film 2.2×10^{-4} cm thick	$\rightarrow 1$ for $T > 125$°C or thicker films, $\eta < 1$	Beeck, Smith, and Wheeler [24, 383]
20. Ni on aluminum oxide	Ethylene hydrogenation	80–140°C	$S_g = 91\ m^2/g$, $V_g = 0.345\ cm^3/g$ $d = 150$ Å	0.2–0.5	<0.5 Calculated $\eta = 0.07$–0.19	Rozovskii, Shchekin, and Pokrovskaya [284], by effect of particle size; calculated values of η assume $\tau = 1$
21. ZnO-Cr_2O_3	$CO + 2H_2 \rightarrow CH_3OH$	280 atm, 330–410°C 0–30% conv. of CO. Gas comp'n = 12% CO, 80% H_2, 8% inert	Assumed $D_{eff} = \theta D/\sqrt{2}$, $\theta = 0.68$	0.5×1.6	0.52–0.95	Pasquon and Dente [246]
22. Dowex-50, cation exchange resin	Hydrolysis of ethyl formate	25°C, dilute solution		0.07	0.62	Smith and Amundsen [325], η estimated by present author. See Section 3.5
23. Amberlite IR-120, cation exchange resin	Inversion of sucrose	50°C, 20% solution		0.024	0.044	Bodamer and Kunin [38]. See Section 3.5

Table 3.1—*Continued*

Catalyst	Reaction	Reaction Conditions	Catalyst Characteristics	Pellet Size (diam in cm)	Effectiveness Factor η	Author and Method
24. Amberlite IR-120, ion exchange resin	Inversion of sucrose	50°C, 20% solution	20-fold variation in degree of crosslinking of polymer	0.045 and somewhat greater	0.25 to 0.0009	Bodamer and Kunin [38]. See Section 3.5
25. Commercial Co-Mo on Al_2O_3	Hydrodesulfurization	350–375°C 50 atm liquid-phase trickle bed	$\bar{d} = 50$ Å	0.5 × 0.5	~0.36	van Deemter [343], by variation of particle size. Reaction rate also increased with reduction in liquid viscosity
26. Dowex 50W × 8 cation exchange resin	Inversion of sucrose	50–70°C	$D_{eff} = 27 \times 10^{-8}$ cm^2/sec at 50°C	0.004 0.077	→1 0.13 (70°C) 0.25 (50°C)	O'Connell [230], by variation of particle size
27. Dowex 50W × 8	Hydration of isobutylene to *t*-butanol	74–97°C	$D_{eff} = 1.7 \times 10^{-5}$ to 5.5×10^{-5} cm^2/sec (varies inversely with temperature)	110–115 mesh mesh 48–65 mesh	0.97 (74°C) 0.60 (94°C) 0.88 (74°C) 0.37 (94°C)	Gupta and Douglas [128], by variation of particle size and analysis
28. 0.5 wt % Pd on alumina, commercial-type pellets	Hydrogenation of α-methylstyrene to cumene (liquid phase)	70–115°C, 1 atm H_2 pressure	$\tau = \frac{D_{1m}\theta}{D_{eff}} = 3.9$	0.32 × 0.32	0.11 (85°C) 0.07 (115°C)	Satterfield, Ma and Sherwood [297], by variation of particle size and analysis
29. 1% Pd on alumina spheres	Hydrogenation of α-methylstyrene to cumene (liquid phase)	50°C, 1 atm H_2 pressure	$\tau = \frac{D_{1m}\theta}{D_{eff}} = 7.5$	0.825	0.006	Satterfield, Pelossof and Sherwood [298], by variation of particle size

of molecular sieves, etc. (Sections 1.5.2 and 1.7.4), and those affecting the local intrinsic activity. The latter may develop from nonuniform distribution of catalyst on a carrier (Section 1.5.1), the nonuniform buildup of carbonaceous deposits or other catalyst poisons (Section 5.1), or by incorporation of molecular sieves into the structure. These problems must be handled on an individual basis, but if sufficient information is available they can frequently be analyzed satisfactorily.

The principal problem in predicting the effective diffusivity given a reasonably isotropic pore structure, stems from the fact that reaction is usually at pressures and temperatures substantially different from those at which measurements have been made. The discussion in Chapter 1 gives guidance on methods of prediction and the following summarizes the conclusions. For gas-phase diffusion by the bulk mode the effects of pressure and temperature have been reasonably well developed. A tortuosity factor found under one set of conditions should apply with little error to other conditions provided that diffusion remains in the bulk mode. If no diffusion measurements are available on the catalyst structure of interest, assumption of a tortuosity factor in the range of 2 to 6 in Equation 1.29 is recommended for the usual catalysts. If gas-phase diffusion is by the Knudsen mode under both reaction conditions and those under which the effective diffusivity by itself has been measured, extrapolation by Equation 1.33 assuming a constant tortuosity factor seems to be satisfactory. In the absence of a diffusion measurement, the assumption of a tortuosity factor in the range of about 2 to 6 is recommended, provided that the pore-size distribution is narrow. Catalysts with a wide pore-size distribution such that much of the diffusion occurs in the transition range provide more of a problem since the relative contributions to the flux from pores of different sizes vary with pressure, temperature, and nature of the diffusing species. The best available procedure at present seems to be to use the parallel-path pore model (Section 1.8.1) for extrapolation, again applying a tortuosity factor in the range of 2 to 6 in the absence of other information. Sections 1.7.3 and 1.7.4 give guidance as to how the possibility of surface diffusion or molecular sieve-type diffusion may be examined. Chapter 1 showed that for bulk diffusion in small pores or transition-range diffusion the effective diffusivity of one gas is a function of the opposing flux of a second gas. Under constant pressure conditions such as those provided in the Wicke-Kallenbach diffusion measurement method, the ratio of the fluxes is inversely proportional to the square root of the molecular weights of the gases. It is important to recognize that *under reaction conditions the flux of reaction products out of a porous catalyst opposing the inward flux of reactant is determined by stoichiometry and not by molecular weight.* For diffusion in the liquid phase, extrapolation to higher temperatures is generally limited by the degree of knowledge concerning bulk diffusion for the system of concern rather than by the effect of the porous structure as such. Relatively little information is available on the effect of temperature on diffusion in the liquid phase and the diffusivity may vary substantially with concentration.

Sufficient studies now exist to verify the close correspondence of theory and experiment when the above complications are not severe. Brief discussion of some of these will help clarify the state of knowledge. Wakao and Smith [356] have developed a method of predicting effectiveness factors using their random-pore model for predicting the effective diffusivity and Smith and his co-workers have tested the theory in a number of experimental studies. The theory involves a microeffectiveness factor relating to possible diffusion limitations through the individual powder particles comprising a catalyst pellet as well as a macroeffectiveness factor relating to the pellet as a whole. In all the studies reported, the microeffectiveness factor was essentially unity, and, indeed, this would be expected to be generally the case except for powder particles containing pores only slightly larger than the diffusing molecules (Section 4.6.2). Predictions of the effective diffusivity by the random-pore model have agreed fairly closely with experiment for pellets compressed from alumina boehmite powder. Two studies of the *o-p* hydrogen conversion on pellets of 7 per cent NiO on alumina [259] and on 25 per cent NiO on alumina [261] compressed to various densities gave agreement within 10 to 15 per cent between experimental and predicted effectiveness factors, which were in the range of 0.15 to 0.30. For the first case about one-half of the diffusion flux was predicted to come via the macropores, and in the second case about 90 per cent. A study of the same reaction on 2 per cent NiO on porous Vycor [260] showed agreement within 10 per cent between experimental and predicted effectiveness factors of about 0.4, using experimentally measured values of the effective diffusivity. At low values of the effectiveness factor, η is not highly sensitive to the effective diffusivity, being proportional to $D_{\text{eff}}^{1/2}$.

In a study by Sterrett and Brown [329] of the same reaction on a ferric oxide gel catalyst of various sizes, η varied from 0.78 to higher values. Theory and experiment could be brought together by a tortuosity factor $\tau_m = 1.4$ as calculated from Equation 1.33 or τ_p of 1.7 as calculated by the parallel-path pore model. These unusually low values of the tortuosity factor suggest that a substantial fraction of the mass transport was by surface diffusion, which seems plausible since all the pores of the catalyst were below about 30 Å radius. The ortho-para hydrogen conversion at about $-196°$ C and 1 atm is a good model reaction since heat effects are negligibly small, the kinetics are relatively simple, and there is no change in number of moles on reaction.

When diffusion is completely by the bulk mode, as in liquid systems, prediction of the effective diffusivity may be easier. Ma studied the hydrogenation of α-methylstyrene to cumene on pellets of Pd/alumina catalyst at 70° to 115°C [297]. The effectiveness factors, found by varying particle size, were from 0.07 to 0.13 on the large pellets and could be brought into agreement with theory using measured values of hydrogen diffusion in α-methylstyrene and cumene and the reasonable tortuosity factor of 3.9. Pelossof [298] studied the same reaction at 50°C on alumina spheres impregnated with palladium and on powdered catalyst. The effectiveness factor was 0.0057 on the spheres. Bringing theory and experiment together required a relatively high tortuosity factor of 7.5.

In an early investigation Johnson, Kreger, and Erickson [159] studied the cracking of gas oil at 480°C and 1 atm on silica-alumina catalysts treated in various ways. On an 0.44-cm-diameter 90 per cent silica–10 per cent alumina commercial bead catalyst, the effectiveness factor was found to be 0.55 by comparing the rates on the bead catalyst and powdered beads for the first 40 per cent of reaction. Theory and experiment could be brought together by a tortuosity factor of about 3.5 assuming the reaction was first order and a Knudsen diffusivity calculated by Equation 1.33. This is remarkably close agreement between theory and experiment considering the uncertainties involved in representing this complex reaction of a wide mixture of substances as a simple first-order process. Diffusion measurements alone in homogeneous silica-alumina catalysts give a tortuosity factor of about 2 to 2.5 (Tables 1.9, 1.11). Higher tortuosity factors were found from the reaction studies on the catalysts treated in various ways, but no information is available on how treatment affected the diffusivity as such.

Weisz and Prater [377] reported an early study of cumene cracking at 420°C and 1 atm pressure on a commercial silica-alumina bead catalyst in various sizes, in which η varied from unity down to 0.15. The effective diffusivity of hydrogen in the catalyst beads at 27°C was measured and extrapolated to reaction conditions using the Knudsen relationship of Equation 1.31. Theory and observation could be brought together by a tortuosity factor of 5.6, although the diffusion of hydrogen alone required a tortuosity factor of 2.3. The biggest source of uncertainty here is probably associated with the complex intrinsic kinetics of the reaction, which was assumed to be simple first order in the analysis. A study by Weisz of burnoff in air of carbon from a silica-alumina cracking catalyst also showed reasonably good agreement between theory and experiment. The analysis is given in Example 4.7.

Certain other experimental studies have shown greater deviation between theory and experiment. Otani and Smith [240] studied the rate of oxidation of carbon monoxide at 275° to 370°C on a 10 per cent nickel oxide-on-alumina catalyst in the form of both powder and $\frac{3}{4}$-in. spherical pellets compressed from the powder. The experimental effectiveness factors, determined by comparison of the rates on pellets and powder, varied from 0.37 to 0.64. Back-calculation gave values of the effective diffusivity 4 to 5 times lower than those that would be predicted by the Wakao and Smith random-pore model. The reason for the discrepancy is not clear. Although the random-pore model gave reasonably good predictions of the effective diffusivity for other pellets pressed from alumina, large spherical pellets such as these may be highly anisotropic [55, 300] or some of the micropores may have been sealed by the pelleting process. Finally, the rate of reaction is greatly inhibited by the product carbon dioxide and rather complex analysis was required to allow for this effect.

3.7 Mass Transfer Limitations to and within Catalyst Particle

In Chapter 2 it was shown that, if the processes of mass transfer and reaction all occur in series, each of these steps may be regarded as a resistance and the observed rate may be set equal to the over-all concentration difference divided by the sum of the resistances. Equation 2.34, for example, shows the formulation for a slurry reactor in which two mass transfer steps and a reaction at the outside surface of catalyst particles each contribute a resistance. As shown in Chapter 2, by proper variation of experimental conditions it is possible to separate the contributions of different steps. When reaction occurs simultaneously with pore diffusion, however, the process cannot be regarded as a case of resistances in series. If both mass transfer to the outside of a porous pellet and pore diffusion are significant resistances, bulk mass transfer coefficients can be determined by the methods of Chapter 2, but the concentration at the outside surface of the catalyst pellet, which is a boundary condition affecting the bulk mass transfer rate, is affected by the internal effectiveness factor. Consider, for example, a spherical pellet of radius R and equate the rate of first-order catalytic reaction within the pellet to the rate of mass transfer to the outside pellet surface:

$$-\frac{1}{V_c}\frac{dn}{dt} = k_v c_s \eta = k_c(c_0 - c_s)\frac{3}{R}. \tag{3.42}$$

Eliminating c_s and rearranging, we obtain

$$-\frac{1}{V_c}\frac{dn}{dt} = \frac{c_0}{(R/3k_c) + (1/k_v \eta)}. \tag{3.43}$$

Although it would be possible from rate data alone to determine the value of the product of k_v and η, it is not possible to determine either separately without additional measurements, e.g., of the effective diffusivity.

With high-area porous catalysts of the sizes and activities typically used in chemical and petroleum processing in packed beds, with gas-phase reaction at typical industrial mass flow rates, and with calculated effectiveness factors approaching unity, mass transfer to the outside surface of the pellet will normally represent an insignificant resistance. If the calculated effectiveness factor is a small fraction of unity, however, it would be wise to determine the possible contributions of bulk mass transfer and thence the proper value of c_s to use in determining η. Conversely, if bulk mass transfer is found to be significant, the effectiveness factor of a high-area porous catalyst would be expected to be very low. These comments apply to systems in which the same phase exists in the pores as in the bulk. If the phases are different or a large distribution coefficient exists between bulk fluid and catalyst, the internal effectiveness factor may approach unity while bulk mass transfer constitutes a significant resistance. This is the case in some reactions catalyzed by ion-exchange resin particles. In laboratory work, mass velocities are frequently low and can lead to significant mass transfer resistance even at relatively low catalyst activities. The above generalizations are intended to be but an approximate guide to expected behavior; analysis of the effects of bulk mass transfer and pore diffusion for any particular case should be made by the quantitative methods described.

3.8 Diffusion Effects on Reaction Intermediates

Some catalysts provide more than one kind of site, and reaction proceeds by consecutive steps on different kinds of sites. These are termed "bifunctional" or "multifunctional" catalysts and are exemplified by the reforming catalysts used for converting petroleum stocks into high octane gasoline. They consist of a highly dispersed metal, usually platinum, on an acidic oxide support, usually an acidified form of alumina. A large variety of reactions occur industrially, but a model reaction which illustrates the key concepts is the isomerization of an *n*-paraffin to an iso-paraffin. A paraffin molecule is dehydrogenated to an olefin on a metal site *a*, the olefin diffuses to an acidic site *b* where it isomerizes, and the isomerized olefin diffuses to a metal site *a* where it undergoes hydrogenation.

$$\text{paraffin} \overset{a}{\rightleftharpoons} \text{olefin} \overset{b}{\rightleftharpoons} \text{iso-olefin} \overset{a}{\rightleftharpoons} \text{iso-paraffin}. \tag{3.44}$$

Both kinds of sites are necessary for any appreciable conversion of paraffin to iso-paraffin. Weisz and Swegler have shown [369, 380] that a mechanical mixture of particles of platinum supported on carbon and particles of an acidic oxide will provide essentially the same yield of iso-paraffin as a dual function catalyst comprising platinum supported on silica-alumina, which is acidic, if the particles in the mechanical mixture are sufficiently fine, but neither type of particle alone forms appreciable product. As particle size of the mechanical mixture is increased, the yield diminishes. From this they conclude that the intermediates are desorbable and are transported from site to site through the gas phase. This further indicates that the rate of the over-all reaction should be fundamentally affected by the thermodynamics of olefin formation since this sets an upper limit on the concentrations of olefins which can be attained. Weisz [368, 369, 371] has developed generalized criteria for proximity requirements for poly-step processes involving desorbable intermediates in terms of the free energy of formation of the intermediate, particle size, effective diffusivity within the catalyst particles, and other variables, and Gunn and Thomas [127] and Thomas and Thomas [334] have presented mathematical analyses of how the distribution of catalyst sites both within catalyst pores and on outer surfaces of catalyst particles may affect the outcome of the reaction. However, some reactions such as dehydrocyclization may involve surface diffusion rather than gas-phase diffusion and the quantitative application of these theoretical considerations requires a fairly detailed knowledge of the true kinetics. Thus, if an olefin intermediate is not desorbed, the catalyst activity cannot be directly related to the olefin concentration in the gas phase. A good review of bifunctional catalysis has been recently published by Sinfelt [322].

Similar poly-step processes occur in noncatalytic reactions. For example, the decomposition of barium carbonate can be greatly enhanced by intimately mixing carbon particles with the powdered carbonate. The mechanism is presumably

$$\begin{aligned} BaCO_3 &\rightleftharpoons BaO + CO_2, \\ CO_2 + C &\rightleftharpoons 2\,CO. \end{aligned} \tag{3.45}$$

The rate of reaction of carbon with CO_2 to form CO can frequently be the rate-limiting step, since it is found that the decomposition rate is greatly affected by the reactivity of the carbon used [294].

4 Diffusion and Reaction in Porous Catalysts II. Complex Cases

4.1 Temperature Gradients

The analyses of the effectiveness factor in Chapter 3 assume the porous structure to be isothermal. Wheeler and others have pointed out that substantial temperature gradients can sometimes occur in practice, and there have been many publications describing mathematical analyses of this effect. A relationship between temperature and concentration which is useful for orientation may be simply derived by considering a boundary surface surrounding some portion of a porous structure. Under steady-state conditions, the diffusion flux of reactants across this boundary surface equals the rate of reaction within the surface. The heat released (or consumed) by the reaction must all be transferred across the same boundary. Hence

$$D_{\text{eff}} \frac{dc}{dx} \cdot \Delta H = \lambda \frac{dT}{dx}, \tag{4.1}$$

or, in integrated form,

$$\Delta T = T - T_s = \frac{(-\Delta H)(D_{\text{eff}})}{\lambda}(c_s - c), \tag{4.2}$$

where T is the temperature at any point within the particle, c (mol/cm^3) is the concentration at the same point, ΔH (cal/mol) is the enthalpy change on reaction, and λ (cal/sec·cm·°K) is the thermal conductivity of the porous solid; T_s and c_s are boundary values, usually at the outer surface of the particle. This relation was first pointed out by Damköhler [87] and shown by Prater [256] to be valid for all kinetics and to apply to any particle geometry.

To obtain temperature (or concentration) as a function of spatial coordinates, the reaction-rate constant is expressed as an Arrhenius function, and the differential equations for simultaneous mass diffusion and heat diffusion are solved, using Equation 4.2 to eliminate one variable. Analytical solution of the differential equations is quite difficult. Beek [25] and Schilson and Amundson [302] used linear approximations to obtain analytical solutions. Tinkler and Pigford [336] applied perturbation techniques to obtain approximate analytical expressions, a procedure that is applicable when temperature gradients are small but not negligible. Petersen [249] presents an asymptotic solution applicable at high values of the Thiele modulus. Other investigators [60, 335, 376] used machine computation methods.

Tinkler and Metzner [335] and Weisz and Hicks [376] give families of curves correlating the effectiveness factor or the maximum temperature rise with other parameters. Tinkler and Metzner treat first- and second-order irreversible reactions in a spherical pellet and treat first-order reaction in a semi-infinite flat plate. Carberry [60] treats first- and second-order reactions in a single pore, which is equivalent mathematically to the semi-infinite plate treatment. Gunn [126] presents analytical expressions for concentration and temperature distributions inside the catalyst particle for first-order irreversible reaction in spherical or flat plate geometry. Hlavàček and Marek [149, 150, 151, 198] treat zero-order reactions in various geometries. Weisz and Hicks treat irreversible first-order reaction in spherical geometry in terms of the Φ_s parameter as well as the Thiele modulus, and their development will be summarized here. The effectiveness factor in all of these cases is defined as the ratio of the actual rate to that which would occur if the pellet interior were all exposed to reactant at the same concentration *and temperature* as that existing at the outside surface of the pellet.

Two new independent parameters must now be introduced over the isothermal case; one is the exponent in the Arrhenius reaction-rate expression, E/RT, and the other a heat generation function

$$c_s(-\Delta H)D_{\text{eff}}/\lambda T_s.$$

Weisz and Hicks symbolize these by γ and β, respectively. Graphs of the effectiveness factor η versus the Thiele modulus ϕ thus involve families of families of curves. The parameter β represents the maximum temperature difference that can exist in the particle relative to the particle surface temperature, $(T - T_s)_{max}/T_s$, under steady-state conditions. This occurs when the concentration drops to essentially zero within the pellet (Equation 4.2, when $c = 0$). (Under transient conditions the maximum temperature difference may exceed that occurring under steady-state conditions [365]. For an exothermic reaction, β is positive. Four η versus ϕ_s graphs, for *first-order irreversible reaction in spheres*, each presenting a family of curves for $0.8 \gtrless \beta \gtrless -0.8$, are given by Weisz and Hicks [376]. These are for values of γ of 10, 20, 30, and 40. Figure 4.1 reproduces the graph for $\gamma = 20$, which represents a fairly typical set of operating variables, e.g., an activation energy of 24 000 cal/g-mol and reaction temperature of 600°K. The curve for $\beta = 0$ represents the isothermal case. Figure 4.1 shows that for exothermic reactions ($\beta > 0$), the effectiveness factor η may exceed unity, since under some sets of circumstances the increase in rate caused by the temperature rise toward the center of the particle more than offsets the decrease in rate caused by the drop in concentration. Thus, the over-all rate of reaction in the particle is greater than it would be if the interior were at the same concentration and temperature as that at the exterior surface. At high values of the Thiele modulus, η becomes inversely proportional to ϕ, as in the isothermal case. Most of the reaction then occurs in a thin shell at the pellet surface, the interior being at a higher temperature than the surface, but essentially isothermal.

The shape of the curves on Figure 4.1 shows that, for highly exothermic reactions at a low value of the Thiele modulus, the value of η is not uniquely defined by β, γ, and ϕ_s. In effect, for one value of ϕ_s there are three different possible values of η corresponding to three different sets of conditions under which the rate of heat production equals the rate of heat removal. The middle one can be shown to be metastable and not realizable in practice. Which of the two remaining sets of conditions will actually be obtained physically depends upon the direction from which the steady-state condition is approached. The fact that either of two rates of heat release can occur is analogous to the ignition situation with an exothermic surface reaction which occurs in the combustion of carbon and which has been observed in the heterogeneous decomposition of hydrogen peroxide on a nonporous active catalyst [110, 299, 351]. In these cases, for example, a momentary disturbance around the metastable point can cause the surface temperature either to rise to a stable level corresponding to a process limited by mass transfer through the laminar layer outside the solid particle or to drop to another stable

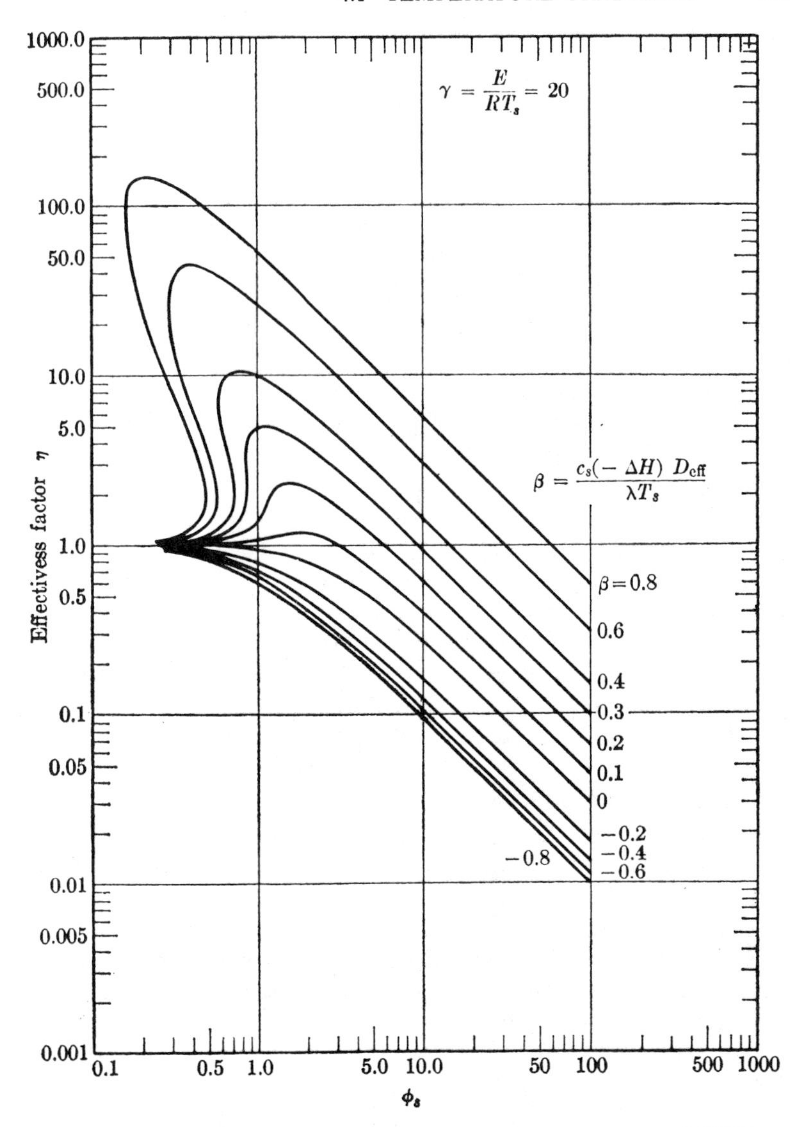

Figure 4.1. Effectiveness factor η as a function of $\phi_s = R\sqrt{k_v/D_{eff}}$ for $\gamma = 20$. First-order reaction in sphere. Weisz and Hicks [376].

level at which the rate of the surface reaction controls. Figures 3.4–3.7, discussed below, make it evident, however, that for porous catalyst pellets the region of multiple solutions corresponds to combinations of large values of β and γ seldom encountered in practice. The conditions under which a unique steady state will occur have been the subject of various mathematical analyses [6, 117, 185, 196, 213, 364].

The abscissa ϕ_s in Figure 4.1 contains the intrinsic rate constant k_v, which is frequently unknown. A more useful method of correlation proceeds by transforming the abscissa to the function Φ_s used previously in the isothermal cases. Figures 3.4–3.7 are graphs of η versus Φ_s for values of $\gamma = 10, 20, 30$, and 40 [376]. The dashed portions of the curves between arrows indicate the regions in which a unique solution does not exist, corresponding to the region of multiple solutions in the η versus ϕ_s curves.

One striking conclusion from these graphs concerns highly exothermic reactions on a catalyst of low thermal conductivity. It is possible to have effectiveness factors greatly exceeding unity at relatively low observed reaction rates or at low values of the Thiele modulus, conditions under which analysis assuming isothermal operation ($\beta = 0$) would indicate an effectiveness factor of essentially unity. The shape of the curves also indicates that small changes in operating conditions or characteristics of the catalyst in this region could cause major changes in the observed reaction rate. Whereas under isothermal conditions the observed activation energy can be as low as one-half the true value when η is sufficiently low (Section 3.3) the opposite effect can occur when $\eta > 1$, i.e., the measured or apparent activation energy can exceed the true value [60]. For kinetic expressions other than first order, such as those of the Langmuir-Hinshelwood type (Equation 4.4), nonisothermal behavior can cause the observed activation energy to even fall below one-half the true value, as shown by Schneider and Mitschka [307]. (See also Section 4.2.2.)

When ϕ_L exceeds about 2 for a first-order reaction, the reactant concentration at the catalyst center approaches zero and the simpler asymptotic solution becomes valid. The number of new independent parameters required for handling nonisothermal conditions can then be reduced from two to one by use of a new parameter equal to $\beta \cdot \gamma$. This new parameter is termed α by Carberry and ε by Tinkler and Metzner, who present a family of curves for various values of α (or ε) on η versus ϕ coordinates for first- and second-order reactions. These show that heat effects becomes less important as reaction order increases, i.e., the family of curves of η versus ϕ for various values of $(\beta \cdot \gamma)$ are more closely compressed for a second-order reaction than for one of first order.

Weekman [362] has analyzed the combined effects of temperature gradients and volume change on reaction when bulk diffusion occurs. For endothermic reactions volume change has little effect on the above analyses. For exothermic reactions, volume contraction can cause a substantial increase in catalyst effectiveness factor and on internal temperatures over that otherwise calculated. Volume expansion works in the opposite direction. For high values of β and γ the variation of β with temperature may be significant and Weekman has presented some analysis of this effect. (See also Section 4.2.2.)

For *endothermic* reactions a good approximation for the effectiveness factor can be obtained by considering only the temperature gradients through the pellet and neglecting concentration gradients, even when these gradients are appreciable. Some analyses by this approach and comparison to the more rigorous method are given by Maymo, Cunningham, and Smith [210].

4.1.1 *Thermal Conductivity of Porous Catalysts*

Whether or not thermal effects can be significant for a particular reaction on a particular catalyst is primarily governed by the value of β. This requires knowledge of the thermal conductivity of porous solids, but relatively little information has been published on the kinds of porous structures used in catalysis. The principal studies appear to be those of Sehr [313], of Mischke and Smith [220], and of Masamune and Smith [203]. The latter two have been further analyzed by Butt [51].

Table 4.1 shows thermal conductivities for several porous oxide catalysts as reported by Sehr [313] at a mean temperature of about 90°C in air at atmospheric pressure. Mischke and Smith [220] measured the thermal conductivity of a series of alumina catalyst pellets in the presence of helium or air at pressures from vacuum to 1 atm. The catalysts were prepared by pelletization of alumina powder under different pressures. Masamune and Smith [203] reported similar studies in the presence of air, carbon dioxide, or helium on a series of metal catalysts prepared by compacting microporous silver powder. Representative results from these two studies are presented in Table 4.2, plus data reported by other workers, usually incidental to other studies. θ_{macro} is the fraction of the porous pellet occupied by pores above about 100 or 120 Å in radius. θ_{micro} is the fraction occupied by smaller pores. $(1 - \theta_{macro} - \theta_{micro})$ is the solid fraction present.

The spread of values of thermal conductivity is remarkably small for the group of nonmetallic substances and does not vary greatly with major differences in void fraction and pore-size distribution, brought

Table 4.1 Thermal Conductivities of Some Porous Catalysts in Air at 1 atm, 90°C [313]

Catalyst	$\lambda_{particle}$ [a] (cal/sec · cm · °C)	λ_{powder} [a] (cal/sec · cm · °C)	Density (g/cm^3)	
			Particle	Powder
Ni-W	1.12×10^{-3}	0.73×10^{-3}	1.83	1.48
Co-Mo de-hydrogenation catalysts	0.83×10^{-3}	0.51×10^{-3}	1.63[b]	1.56[b]
	0.58×10^{-3}	0.33×10^{-3}	1.54[c]	1.09[c]
Chromia-alumina reforming catalyst	0.70×10^{-3}	0.42×10^{-3}	1.4	1.06
Silica-alumina cracking catalyst	0.86×10^{-3}	0.43×10^{-3}	1.25	0.82
Pt-alumina reforming catalyst	0.53×10^{-3}	0.31×10^{-3}	1.15	0.88
Activated carbon	0.64×10^{-3}	0.40×10^{-3}	0.65	0.52

[a] Multiply values in these columns by 241.9 to convert to units of Btu/hr·ft·°F.
[b] 3.6% CoO and 7.1% MoO_3 on α-alumina, 180 m^2/g.
[c] 3.4% CoO and 11.3% MoO_3 on β-alumina, 128 m^2/g.

about by varying the pelletizing pressure (void fractions were not reported by Sehr). The additional fact that the thermal conductivity under vacuum of compacted silver powder is essentially the same as that of alumina under vacuum emphasizes that this property is dependent primarily upon geometrical considerations rather than the thermal conductivity of the solid as such. This is further brought out by a summary published by Masamune and Smith [202] of studies of the thermal conductivity of beds of spherical particles of a wide range of metallic and nonmetallic substances (silicon carbide, glass beads, quartz, steel balls, steel shot, lead, glass, magnesium oxide). The ratio of thermal conductivity of the porous bed of solid under vacuum to that of the nonporous solid itself varied from about 0.05 to 0.001, although the void fractions of the porous beds were all within the range of 0.27 to 0.45. There was no significant correlation between conductivity of the solid and the conductivity of the bed. For sintered Cu-Sn alloy particles at 1 atm pressure the ratio varied from about 0.02 to 0.07 at void fractions of 0.33 to 0.4 to about 0.13 at a void fraction of 0.2.

The last two groups of studies in Table 4.2 are representative of the various investigations with beds of spherical particles. It is evident

Table 4.2 Thermal Conductivities of Some Porous Materials

Substance	Fluid in Pores	Temp (°C)	Pellet Density (g/cm³)	θ_{macro}	θ_{micro}	λ(cal/sec · cm · °C) 1 atm	vacuum	Reference
Alumina (boehmite) pellets	air	50	1.12 0.67	0.134 0.450	0.409 0.275	5.2×10^{-4} 3×10^{-4}	3.9×10^{-4} 1.65×10^{-4}	Mischke and Smith [220]
Pellets of silver powder	air	34	2.96 1.35	0.144 0.61	0.574 0.261	17×10^{-4} 4.0×10^{-4}	1.5×10^{-4} 2.2×10^{-4}	Masamune and Smith [203]
Cu on MgO, pellets[a]	air (?)	25–170	0.7 to 1.20[a]			1.8×10^{-4} to 4×10^{-4}		Cunningham, Carberry, and Smith [81]
Pt on alumina pellets, 0.05 wt %	air	—	1.34	0.35	0.15	3.5×10^{-4}		Miller and Deans [218]
Pt on alumina pellets	H_2	68	0.57	0.56	0.23	6.2×10^{-4}		Maymo and Smith [211]
Bed of stainless steel shot, 71 μ in diameter	air	42	5.77 (bed density)	0.264	—	6.2×10^{-4}	4.1×10^{-5}	Masamune and Smith [202]
Bed of glass beads 29, 80, 200 or 470 μ in diameter	air	42	1.50 (bed density)	0.38	—	4.4×10^{-4}	1.24×10^{-4}	Masamune and Smith [202]

[a] Carbon deposits present to various degrees.

that for these beds conductivity under vacuum is dependent primarily on the area of contact between particles and would be expected to be greater for rough than for smooth surfaces.

Even the most dense silver pellet in Table 4.2 exhibited a thermal conductivity but a small fraction of that of solid silver (which is about 1 cal/sec·cm·°C). If porous metals were prepared by a sintering process, however, as by the usual powder metallurgy methods, one anticipates that the ratio of thermal conductivity to that for pure solid would be substantially greater, as indicated by the results with Cu-Sn alloy particles.

Both the solid and fluid phases are continuous in porous catalysts, so the thermal conductivity may be modeled as two conducting paths in parallel with transfer of heat between the two. The latter contribution is particularly complicated since the pore-size distribution of many catalysts is such that at atmospheric pressure diffusion is in the transition region between the Knudsen and bulk modes where the thermal conductivity of a gas varies significantly with pore size or pressure. The prediction of a model will vary with the geometry postulated for the two paths, but it is evident that the effective conductivity will be determined primarily by that of the phase with the greater thermal conductivity. The greater the differences between the thermal conductivities of the two phases, the greater the divergence between the predictions of different models. Butt [51] has developed a model for catalyst pellets which shows good agreement with the data of Smith and co-workers [203, 220]. Krupiczka [178] gives several examples of models with particular reference to beds of granular materials and gives an extensive compilation of experimental data. Models have also been developed for such structures as sandstone, limestone, foamed glass, and foamed polystyrene [121, 181] and for beds of spherical particles [202]. However, models developed for completely enclosed pores should not be used for porous catalysts.

Some feeling for magnitude is helpful at this point. Reaction conditions will usually be at pressures of 1 atm or higher and at higher temperatures than those at which the measurements reported in Tables 4.1 and 4.2 have been made. The thermal conductivity of a gas is almost independent of pressure when the mean free path is substantially less than the pore size. At room temperature the thermal conductivity of air is about 6×10^{-5} cal/sec·cm·°C, that of hydrogen is about 4.2×10^{-4} and that of a wide range of organic vapors, polar and nonpolar, varies from about 2×10^{-5} to 6×10^{-5}. Omitting hydrogen (and helium), these are an order of magnitude less than that of porous catalysts under vacuum. Thermal conductivities of simple organic liquids are usually 10 to 100 times greater than that of the vapor at the same temperature. Typical values for nonpolar liquids at room temperature are in the range of 2×10^{-4} to 5×10^{-4} cal/cm·sec·°C, and are two or three times greater than this for highly polar substances.

Over the temperature range of about 60–200°C, the thermal conductivity of a gas typically increases about linearly with temperature and doubles over about a 200°C temperature range. Data in Table 4.1 and 4.2 are for catalyst pores filled with air at 1 atm, with one exception. The thermal conductivity of a liquid or vapor under reaction conditions will usually be greater than that of air at room temperature, so the values reported for effective thermal conductivity of porous catalysts in air usually represent the minimum values to be expected during reaction. Temperature gradients within a catalyst particle calculated by their use would thus represent probable maximum values.

It is clear that in catalyst operation the greatest deviations from isothermal behavior will occur with high effective diffusivity in the catalyst, low thermal conductivity, and high heat of reaction. Increasing the pressure will increase the degree of departure of a gaseous reaction from isothermal behavior, if diffusion is initially in the Knudsen or transition region, since the increased pressure increases the diffusion flux. It will generally be found that even with catalysts of thermal conductivity as low as that of silica-alumina, essentially isothermal operation need not be questioned for reaction at atmospheric pressure or less unless (a) a substantial fraction of the porosity is represented by pores so large that bulk diffusion occurs in them, or (b) the heat of reaction is unusually high (e.g., over 40 kcal/g-mol). However, since the interaction of many variables determines the degree of departure from nonisothermal operation, the best procedure is to estimate the value of β applicable to each specific situation. The illustrative examples below will provide additional orientation.

Example 4.1 Estimation of Temperature Effect on Effectiveness Factor η

Consider the data on catalytic cracking of gas oil on a silica-alumina catalyst as analyzed in Example 3.1. The endothermic heat of reaction varies with degree of reaction because of secondary processes, but the maximum is about 40 000 cal/g-mol. Assume the thermal conductivity of silica-alumina to be that given by Sehr [313] (see Table 4.1). The effective diffusivity was estimated in the earlier example to be about 8.0×10^{-4} cm^2/sec. Consider conditions at the entrance to the bed, with a reaction temperature of 630°C.

$$\beta = \frac{c_s(-\Delta H)D_{eff}}{\lambda T_s} = \frac{\left(\dfrac{1}{22\,400} \times \dfrac{273}{903}\right)(-40\,000)(8.0 \times 10^{-4})}{(8.6 \times 10^{-4})(903)}$$

$$= -0.00052.$$

Since $\beta = \Delta T_{max}/T_s$, the maximum temperature difference that could exist between particle surface and interior under any conditions is about 0.5°C.

The maximum activation energy for this reaction is about 40 000 cal/g-mol. Hence, $\gamma = 40\,000/(2)(903) = 22$. Figure 3.5 or Figure 4.1 for $\gamma = 20$ shows that the curve for $\beta = -0.0005$ would not be perceptibly different from that for $\beta = 0$. Hence the assumption of isothermal operation here is excellent.

Example 4.2 Estimation of Temperature Effect on Effectiveness Factor η

Consider the dehydrogenation of cyclohexane over a platinum-alumina reforming catalyst at 25 atm pressure and 450°C. A large excess of hydrogen is used to prevent carbon formation on the catalyst, so a 4:1 hydrogen/hydrocarbon ratio will be considered. Prater [256] gives the following data:

ΔH of reaction $= +\,52.5 \times 10^3$ cal/mole;

$D_{eff} = 16 \times 10^{-3}$ cm/sec² for cyclohexane (assume this is completely Knudsen diffusion);

thermal conductivity $= 5.3 \times 10^{-4}$ cal/sec · cm · °C.

At 450°C, the reaction will be effectively irreversible. The value of β is

$$\frac{\left(\dfrac{1}{22\,400} \times \dfrac{273}{723} \times \dfrac{25}{5}\right)(-52\,500)(16 \times 10^{-3})}{(5.3 \times 10^{-4})(723)} = -0.18.$$

The reaction is highly endothermic, so it is reasonable to assume that the activation energy is approximately equal to the enthalpy change of the reaction. Thus $\gamma = 52\,500/(2)(723) = 36$. Figure 3.7 of η versus Φ_s for $\gamma = 40$ shows that, under highly diffusion-limiting cases, η may be as small as one-quarter of that for isothermal conditions; i.e., the observed rate could be as low as one-quarter of that obtainable under the same conditions if the same catalyst had a high thermal conductivity. The temperature in the center of a catalyst pellet could be as much as $(0.18)(450 + 273) = 130$°C below that at the outside surface.

In this particular reaction, the equilibrium composition shifts rapidly with temperature, and at 400°C instead of 450°C, for example, a substantial fraction of cyclohexane would remain at equilibrium. Figures 3.4 through 3.7 and 3.9

assume reaction to be irreversible. Methods of analyzing reversible reactions are discussed in Section 4.4.

4.1.2 *Tests of Theory, Nonisothermal Reaction*

Two very similar studies, by Maymo and Smith [211] and by Miller and Deans [218], have provided a good test of the theory in the presence of temperature gradients within a catalyst particle. Both studied the reaction of oxygen in excess hydrogen on platinum-on-alumina catalysts of such large pore sizes that diffusion was essentially by the bulk mode. In the Maymo and Smith study, intrinsic kinetics were determined on catalyst powder; the activity of a pellet compressed from the powder was varied by changing the ratio of catalyst powder to inert alumina in the mixture compressed to form a pellet. A single 1.86-cm-diameter pellet was studied. With the most active pellet, temperature differences of as much as 115°C occurred between bulk gas and outside pellet surface, and temperature differences between outside surface and pellet center were as much as 300°C. Indeed, the local surface temperature varied with position around the pellet (a sphere is not a "uniformly accessible" geometry unless it is surrounded by a stagant fluid, and the point heat transfer coefficient would be expected to vary with position). The intrinsic kinetic study showed the rate to be proportional to the 0.8 power of the oxygen and to have an apparent activation energy of 5.2 kcal. The thermal conductivity of the pellets was measured, from which the effective diffusivity was calculated, using Equation 4.2. Effectiveness factors were predicted by numerically integrating the mathematical expressions for simultaneous diffusion and reaction, using the intrinsic kinetic expression and the above effective diffusivities. Omitting the most rapid reaction studies, in which mass transfer to the outside surface of the pellet was highly important, the predicted effectiveness factors, which varied from about 0.5 to 1.4, averaged within about 7 per cent of those found experimentally. (Temperature differences between center and outside of the pellet surface as found experimentally varied from 8–102°C for this group of the studies.)

In the Miller and Deans study [218], commercial platinized alumina cylinders were used, the effective diffusivity and thermal conductivity were directly measured, and a rather complex procedure was used to isolate the intrinsic kinetics from data obtained on the cylinders. The final equation, representing the rate as proportional to the 0.8 power of the oxygen with an activation energy of 5.5 kcal, agreed closely with that found by Maymo and Smith for the same type of catalyst [211]. Their experimental effectiveness factors were obtained by a somewhat indirect procedure and varied from 0.53 to 0.90. Those calculated assuming isothermal operation were from 10 to 40 per cent below these values. The experimental values of η were in turn 20 to 30 per cent below those predicted by the Weisz and Hicks method. However, the kinetic expression, the geometry, and the nonequimolar counterdiffusion flux all differ from those used in the Weisz and Hicks [376] analysis. Observed temperature differences between center and outside pellet surface, from 60–30°C, agreed within an average of about 10 per cent with those calculated, and these in turn were from 60–90 per cent of the maximum temperature difference as calculated by an equivalent of Equation 4.2 for the center concentration, $c_i = 0$.

A study of the hydrogenation of ethylene (in a 17 per cent ethylene, 83 per cent hydrogen mixture) at 1 atm on a copper-magnesium oxide catalyst [81] on $\frac{1}{2}$-in. spherical pellets at temperatures up to about 130°C showed temperature differences up to 20°C between pellet surface and pellet center. A study of the hydrogenation of benzene vapor on a single $\frac{1}{2}$-in. cylindrical pellet comprising 50 per cent nickel on kieselguhr at temperatures up to 133°C [158] also showed substantial temperature gradients from the inlet to the exit of the reactor, from bulk gas to pellet surface, as well as through the pellet. Both reactions are highly exothermic and rapid, complex kinetically, and theoretical interpretation was uncertain. A theoretical analysis by Bischoff [35] gives the temperature distribution expected for diffusion-limited reaction in a sphere when the surface temperature varies linearly with position on the circumference.

4.2 Complex Irreversible Kinetic Expressions

4.2.1 *Isothermal Reaction*

The mathematical analyses discussed in the previous sections of Chapter 3 apply to a single reactant, an irreversible reaction, and a simple power-law expression for the rate of reaction. We now consider methods of estimating the effectiveness factor for more complex kinetic expressions, as represented, for example, by a Langmuir-Hinshelwood (Hougen-Watson) type of rate equation. A first approach to developing a generalized method was published by Chu and Hougen, who were particularly concerned with the oxidation of NO to NO_2 [72]. Several investigators have obtained closed form solutions by making various assumptions to simplify the mathematical treatment [3, 284]. The specific case of cracking of cumene to benzene and propylene has been analyzed by Prater and Lago using numerical techniques [257]. A series of papers by Schneider and Mitschka [303–306] present the generalized results of numerical analyses for isothermal reaction in an infinite flat plate (infinite slab) for Langmuir-Hinshelwood type kinetics and for both reversible and irreversible reactions. The presentation that follows here is a generalized method developed by Roberts [272, 273] on the basis of numerical computations which has the practical advantage of relating η to the modulus Φ_L, which contains only quantities that can be observed or predicted.

The general chemical equation describing the reactions under consideration is

$$\mathrm{A} + b\mathrm{B} + \cdots \rightarrow x\mathrm{X} + y\mathrm{Y} + \cdots. \tag{4.3}$$

Two types of kinetic expressions will be considered.

Type I

The rate equation is taken to be

$$kp_{\mathrm{A}}/(1 + K_{\mathrm{A}}\, p_{\mathrm{A}} + \sum_i K_i\, p_i) = -\frac{1}{V}\frac{dn_{\mathrm{A}}}{dt}, \tag{4.4}$$

where index i is used to denote any reaction product or reactant other than A. This expression includes reactions in which A decomposes or isomerizes by a first-order process, or reaction of A with B in which the concentration of B does not appear in the numerator, but may appear in the denominator. For the reaction of A with B, such an expression might result, for example, if adsorption of A on the catalyst is the rate-controlling process. It can be derived by assuming that the rate-limiting step is either the surface reaction or the adsorption of a single reactant without dissociation and allows for possible inhibition of the reaction rate by either reactants or products.

Type II

The reaction rate is assumed to obey the equation

$$k_2\, p_{\mathrm{A}}\, p_{\mathrm{B}}/(1 + K_{\mathrm{A}}\, p_{\mathrm{A}} + K_{\mathrm{B}}\, p_{\mathrm{B}} + K_X\, p_X + \cdots)^2 = -\frac{1}{V}\frac{dn_{\mathrm{A}}}{dt}. \tag{4.5}$$

When A is the only reactant, p_{B} is replaced by p_{A} in Equation 4.5 and $K_{\mathrm{B}}p_{\mathrm{B}}$ disappears from the denominator. This expression thus includes the case of two or more molecules of A reacting with one another, in which case B = A.

For both types of kinetic expressions it is assumed that the catalyst mass is infinite in two directions and of thickness L in the third, and is exposed to a reactant gas on one face and sealed on the other. It is further assumed that the pellet is isothermal, and that the ideal gas laws are applicable. Diffusion is assumed to obey Fick's law and the diffusivities of all species are taken to be constant but not necessarily equal. Diffusion of each component is assumed to be unaffected by the flux of other species, so the total pressure in the catalyst interior may be substantially different than that at the outside surface (see Section 4.5). For gases, the assumption of a constant effective diffusivity is justified for Knudsen diffusion, for binary, bulk equimolal counterdiffusion or if one component of the gas mixture is present in great excess. This major component might be a reactant, such as hydrogen in a hydrogenation reaction or a "nonparticipating" component such as steam in a dehydrogenation reaction or an inert gas in the removal of carbonaceous deposits by oxidation. Equations 1.15 or 1.34 may be used to estimate whether a component is present in sufficient excess to justify the assumption of a constant effective diffusivity.

For Type I kinetics, determination of the effectiveness factor requires the use of one additional modulus, $Kp_{A,s}$ where $p_{A,s}$ is the partial pressure of reactant A at the outside surface of the pellet and K is given by

$$K = [K_A - D_A \sum_i (K_i v_i/D_i)]/\omega \tag{4.6}$$

and

$$\omega = 1 + \sum_i K_i[p_{i,s} + (p_{A,s} v_i D_A/D_i)]. \tag{4.7}$$

The stoichiometric coefficient v_i includes all species other than A. It is negative for a reactant other than A and positive for all products (i.e., in Equation 4.3, b is negative and x and y are positive). The value of ω will normally be positive, but in the case of a reaction having a reactant in addition to A, a negative value of ω could result if the second reactant had a very large value of K_i and a very small value of Dp_s/v. The method presented here cannot be used for negative values of ω. Since ω is dimensionless, K has the dimensions of an adsorption constant. As the values of K_A and the various K_i become so small that the reaction approaches simple first order, K approaches zero; a negative value of K indicates that the sum of the groups KvD_A/D_i for the products is greater than that for the reactants. Qualitatively, a negative value of K indicates inhibition by reaction products.

Figure 4.2 is a graph of η versus Φ_L, for various values of $Kp_{A,s}$ and for first-, second-, and zero-order reactions. The curve for zero-order reaction corresponds to a value of $Kp_{A,s}$ approaching infinity and that for first-order, to a value approaching zero. That for a second-order expression, however, cuts across the family of curves. A negative value of $Kp_{A,s}$ indicates significant product adsorption effects, as noted above. The minimum possible value of $Kp_{A,s}$ is -1. Using this plot, η can be determined directly from the values of Φ_L and $Kp_{A,s}$. The error involved in the use of an integer-power approximation can also be estimated for any case from the figure. The use of this plot is illustrated in Example 4.3.

Example 4.3 Estimation of Effectiveness Factor η; Isothermal, Complex Irreversible Reaction

The reaction of carbon dioxide with solid carbon is retarded by carbon monoxide, and a complex kinetic expression is required to represent the rate. Walker, Rusinko, and Austin [358] studied the reaction of carbon dioxide with spectroscopic carbon, a finely porous material, at temperatures ranging from 950–1305°C, and at various carbon dioxide partial pressures. The mathematical relationships for this reaction are the same as those for decomposition or isomerization of a single reactant on a porous catalyst, except that the porosity and, hence, the effective diffusivity will increase as reaction proceeds. The reaction-rate data of Walker *et al.*, however, are only for the first 11 per cent of reaction, so the change in diffusivity during a run is relatively small.

Starting with a single cylindrical shape weighing 8.8 g initially, at a carbon dioxide partial pressure of 0.75 atm, and a temperature of 1000°C, the rate of reaction was 0.125 g of carbon per hour. Presumably, the partial pressure of carbon monoxide at the exterior of the carbon was zero during the run; and it will be assumed that nitrogen, which was present in the feed stream, does not enter into the rate equation. The average rate of reaction during consumption of the first 11 per cent of the carbon was $0.125/(8.8 \times 0.945) = 0.015$ g C/g C · hr. The average void fraction (cm^3/cm^3) of this sample during the period of the burnoff was about 0.36. Taking 2.27 as the true density (ρ_t) of carbon, the apparent density of the particle is then

$$\rho_p = (1 - \theta)\rho_t = (0.64)(2.27) = 1.45\,g/cm^3,$$

$$-\frac{1}{V_c}\frac{dn}{dt} = \frac{(0.015)}{12}\frac{(1.45)}{3600} = 5.04 \times 10^{-7}\,mol/cm^3 \cdot sec.$$

The effective diffusivity $D_{eff} = 0.013$ cm² sec at normal temperature (294°K) and pressure. Diffusion apparently occurred in the transition region beween

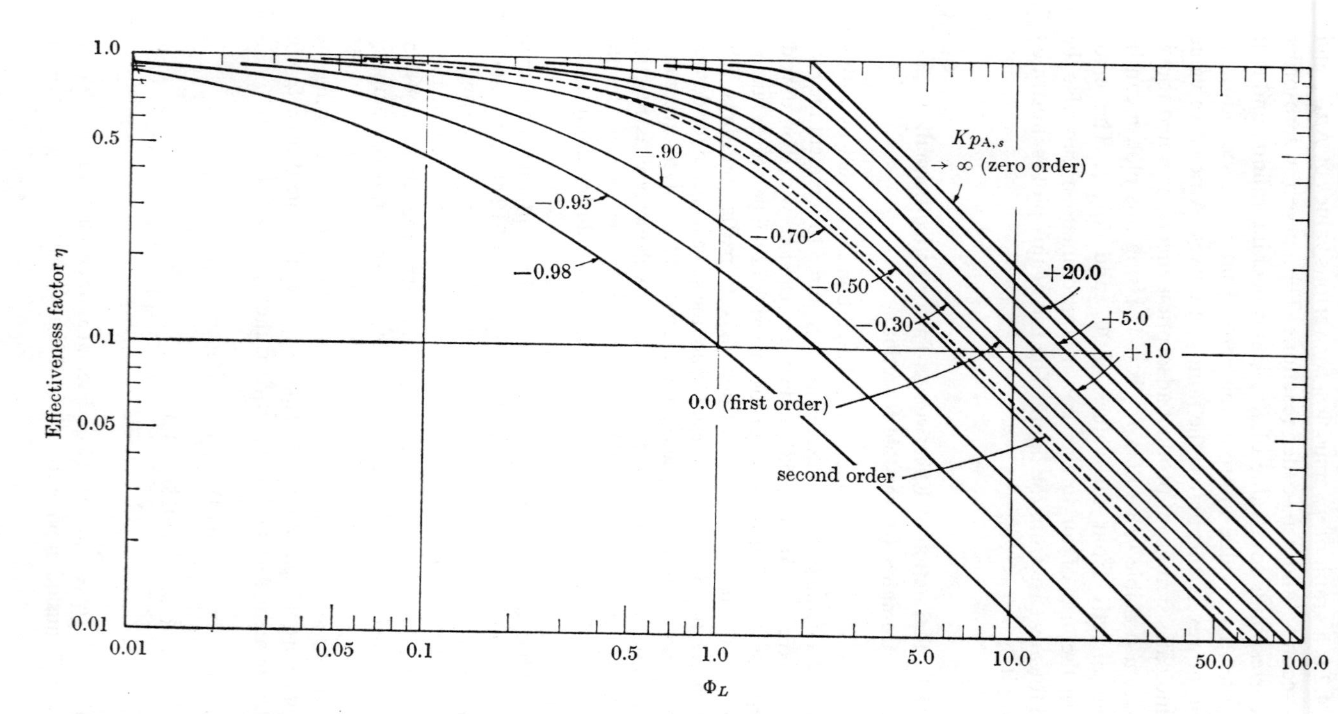

Figure 4.2. Effectiveness factor η as a function of Φ_L (Equation 3.32). Type I kinetics in flat plate geometry (Equation 4.4). Roberts and Satterfield [272].

Knudsen and bulk diffusion. For carbon dioxide counterdiffusing through helium in a similar graphite electrode between 30° and about 400°C at a total pressure of 1 atm, Nichols [226] reported the temperature exponent to be about 0.98. Using this value,

$$D_{\text{eff}} \text{ at } 1000°\text{C} = 0.013 \left(\frac{1273}{294}\right)^{0.98} = 0.0545 \text{ cm}^2/\text{sec}.$$

The external concentration of carbon dioxide is

$$c_{A,s} = 0.75/(82.06)(1273) = 7.16 \times 10^{-6} \text{ mol/cm}^3.$$

The dimension L, the ratio of volume to surface for the cylinder of carbon, was about 0.298 cm.

$$\Phi_L = \frac{L^2}{DC_{A,s}}\left(\frac{-1}{V_c}\frac{dn}{dt}\right) = \frac{(0.298)^2(5.04 \times 10^{-7})}{(7.6 \times 10^{-6})(0.0545)},$$

$$\Phi_L = 0.108.$$

The authors did not determine a rate equation for the reaction, but several other investigators have reported that on each of several types of carbon it is of the form

$$r = kp_{CO_2}/(1 + K_{CO}p_{CO} + K_{CO_2}p_{CO_2}).$$

Wu [393] reports values of the constants for electrode carbon over the temperature and pressure range of interest, from which we estimate that $K_{CO_2} = 2.4$ atm^{-1}, and $K_{CO} = 63$ atm^{-1}. Assuming that these can be applied to the present case,

$$K = (K_A - D_A \sum K_i \nu_i/D_i)/\omega = \left[K_{CO_2} - \frac{D_{CO_2}}{D_{CO}} K_{CO}\,\nu_{CO}\right]/\omega.$$

The diffusivity is approximately inversely proportional to the square root of molecular weight.

$$(D_{CO_2}/D_{CO}) = \sqrt{28/44} = 0.80,$$

$$K = (2.4 - 0.80 \times 63 \times 2)/\omega = -99/\omega,$$

$$\omega = 1 + \sum K_i[p_{i,s} + (p_{A,s} D_A \nu_i/D_i)]$$

$$= 1 + \left(\frac{D_{CO_2}}{D_{CO}}\right) \times p_{CO_2,s} K_{CO}\,\nu_{CO} = 1 + 0.80 \times 0.75 \times 63 \times 2 = 77$$

(note that $p_{i,s} = p_{CO,s} = 0$),

$$Kp_{A,s} = -(99/77) \times 0.75 = -0.965.$$

Interpolating between the curves for $Kp_{A,s} = -0.95$ and -0.98 in Figure 4.2, the effectiveness factor η for this run is about 0.5. Therefore, internal diffusion effects are predicted to be significant. This conclusion is confirmed by the nonuniform porosity profiles after reaction reported by Walker and co-workers. (A theoretical analysis of their data was subsequently published by Austin and Walker [14].) Since K_{CO} increases with decreasing temperature more rapidly than does K_{CO_2}, the retarding effect of CO would be expected to be even more significant at temperatures below 1000°C.

From the intrinsic reaction rate data on electrode carbon published by Reif [265] and taking $p_{A,s} = 0.75$, the value of $Kp_{A,s} = -0.970$, which is very close to that calculated from Wu's data. Actually, Wu's correlation for coal coke gives at this temperature a value of $Kp_{A,s} = -0.965$, identical to the value for electrode carbon. For the same observed rate of reaction, the effectiveness factors for these two types of carbon would, therefore, be very close to the value calculated above. More details of this calculation have been published [272].

Note that, if the reaction were assumed to be of simple first order, the effectiveness factor would erroneously be calculated to be nearly unity — i.e., diffusional effects would be thought to be insignificant. Even if a simple second-order reaction were assumed, the effectiveness factor would be taken to be about 0.92. The use of charts for Φ_L instead of $\frac{1}{9}\Phi_s$ makes little difference (see Section 4.3).

The same reaction was studied earlier by Wicke and Hedden [386] at temperatures of 1000–1100°C and carbon dioxide partial pressures of about 0.1–0.7 atm. They demonstrated that pore diffusion was highly significant for their system by two kinds of results: (1) At a fixed temperature of 1074°C and fixed partial pressure of carbon dioxide of 124 Torr the rate of reaction in the presence of carbon dioxide alone exceeded that of a mixture of carbon dioxide and helium, which exceeded that of a mixture of carbon dioxide and nitrogen. (2) The apparent activation energy was about 43 kcal/g-mol versus values of about 85 reported by other investigators.

For Type II kinetics, an additional parameter $\boldsymbol{E}$ is required in addition to $Kp_{A,s}$ and Φ_L. K and ω are defined as for Type I kinetics, where i is any species other than A and the stoichiometric coefficient ν is taken to be negative for a reactant and positive for a product; i.e., in Equation 4.3, b is negative and x and y are positive. $\boldsymbol{E}$ is defined by

$$\boldsymbol{E} \equiv \left(\frac{-D_B\, p_{B,s}}{\nu_B D_A\, p_{A,s}}\right) - 1. \tag{4.8}$$

For any real system, A is chosen to designate that reactant which permits $\boldsymbol{E}$ to be zero or positive. It is seen that $\boldsymbol{E}$ will have a large value when B is present in considerable excess over A and/or has a substantially higher effective diffusivity than A. It may be thought of as a "modified stoichiometric excess."

Figures 4.3 and 4.4 are plots of η versus Φ_L for selected values of $Kp_{A,s}$. For Figure 4.3, $\boldsymbol{E} = 0$; for Figure 4.4, $\boldsymbol{E} = 10$. A plot for $\boldsymbol{E} = 1$ has been published [273] as well as plots of η versus a modified Thiele modulus for values of $\boldsymbol{E} = 0$, 1, and 10. For orientation, the zero-order curve is shown in Figures 4.3 and 4.4 and the second-order curve for $\boldsymbol{E} = 0$ in Figure 4.3.

Two characteristics of the curves in Figure 4.4 deserve comment. In the first place, effectiveness factors greater than unity result over a range of values of Φ_L when $Kp_{A,s} = 10$ or 100. This is a consequence of the fact that the rate equation, Equation 4.5, possesses a maximum under certain conditions. In the Langmuir-Hinshelwood model, the reaction rate is proportional to the product of the concentrations of adsorbed A and adsorbed B, but the two reactants compete for sites on the catalyst surface. If A is strongly adsorbed relative to B, an increase in p_A at constant p_B will displace B from catalytic sites and this can cause the reaction rate to decrease. It can be shown that effectiveness factors greater than unity will result when $Kp_{A,s}$ is greater than $(\boldsymbol{E} + 2)/\boldsymbol{E}$.

Secondly, for such combinations as $\boldsymbol{E} = 10$, $Kp_{A,s} = 10$ or 100, and for $\boldsymbol{E} = 1$, $Kp_{A,s} = 100$, a range exists for which η is a multiple-valued function of the Thiele modulus, analogous to that found for nonisothermal systems obeying simple kinetics discussed in Section 3.7. Here such effects are also predicted to be possible for *isothermal* reactions and complex kinetics. These regions, represented by dashed lines on the figures, represent unstable operation; the steady-state reaction rate cannot be uniquely determined by specifying the conditions outside the catalyst pellets. The direction from which steady state is approached may instead determine which effectiveness factor is eventually realized. Luss and Amundson [197] have investigated the kinds of kinetic expressions for which a multiple state solution is impossible. The effect of geometry on predicted effectiveness factor is probably also much greater in this region than in most others.

The $\eta - \Phi_L$ curve for Type II kinetics always lies below the curve for Type I kinetics, if $Kp_{A,s}$ is the same in both cases and is negative. This is illustrated by the dashed curve in Figure 4.3 for Type I kinetics, $Kp_{A,s} = -0.90$.

Figure 4.4 also shows that when $Kp_{A,s}$ is large and $\boldsymbol{E}$ is greater than zero, the $\eta - \Phi_L$ curve lies above that for a zero-order reaction. However, in these cases, the diffusional retardation persists to lower values of Φ_L than it does for a zero-order reaction.

Figure 4.5 is a crossplot that illustrates the effect of the modified stoichiometric excess $\boldsymbol{E}$ on the effectiveness factor at constant values

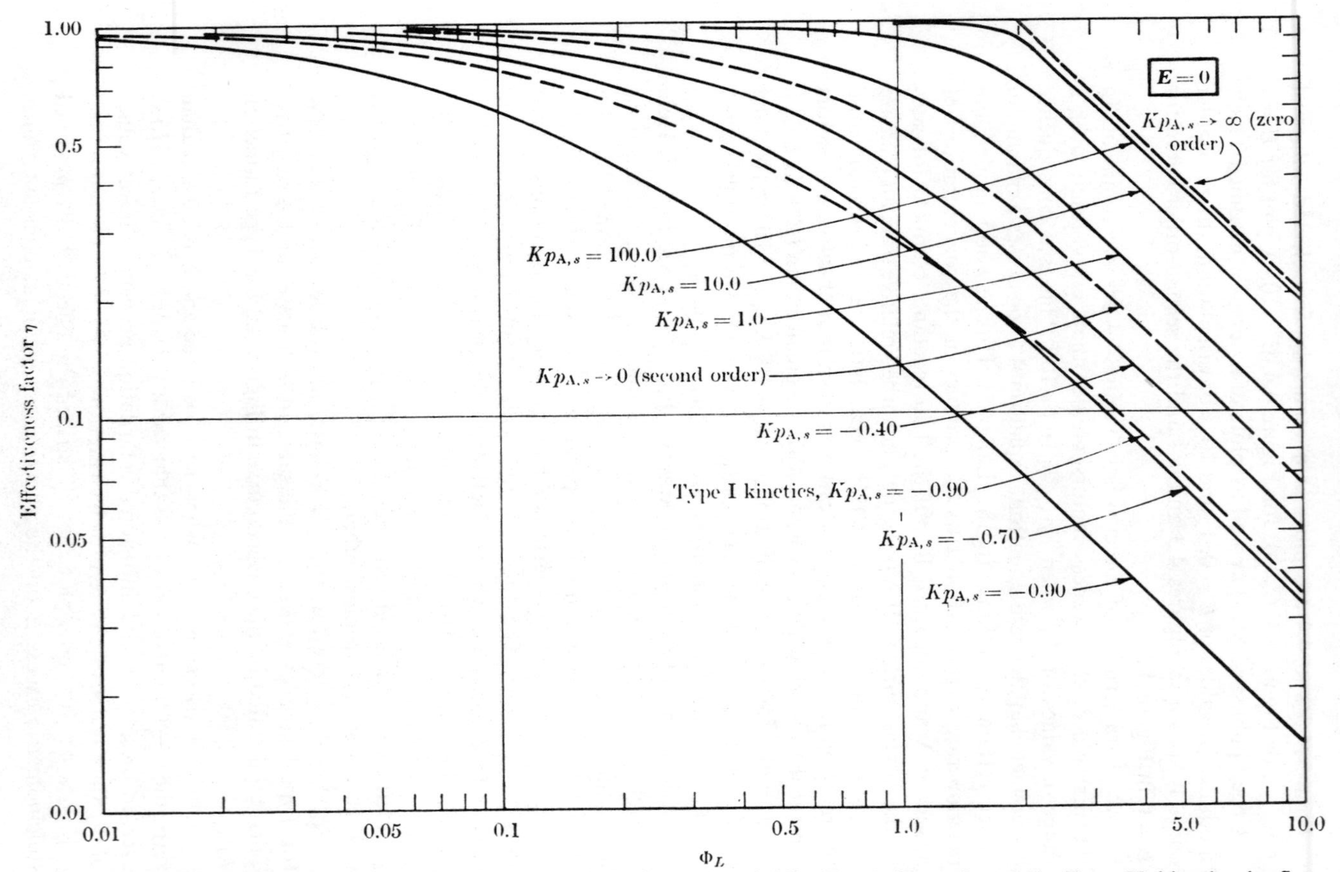

Figure 4.3. Effectiveness factor η as a function of Φ_L (Equation 3.32). $E = 0$ (Equation 4.8). Type II kinetics in flat plate geometry (Equation 4.5). Roberts and Satterfield [273].

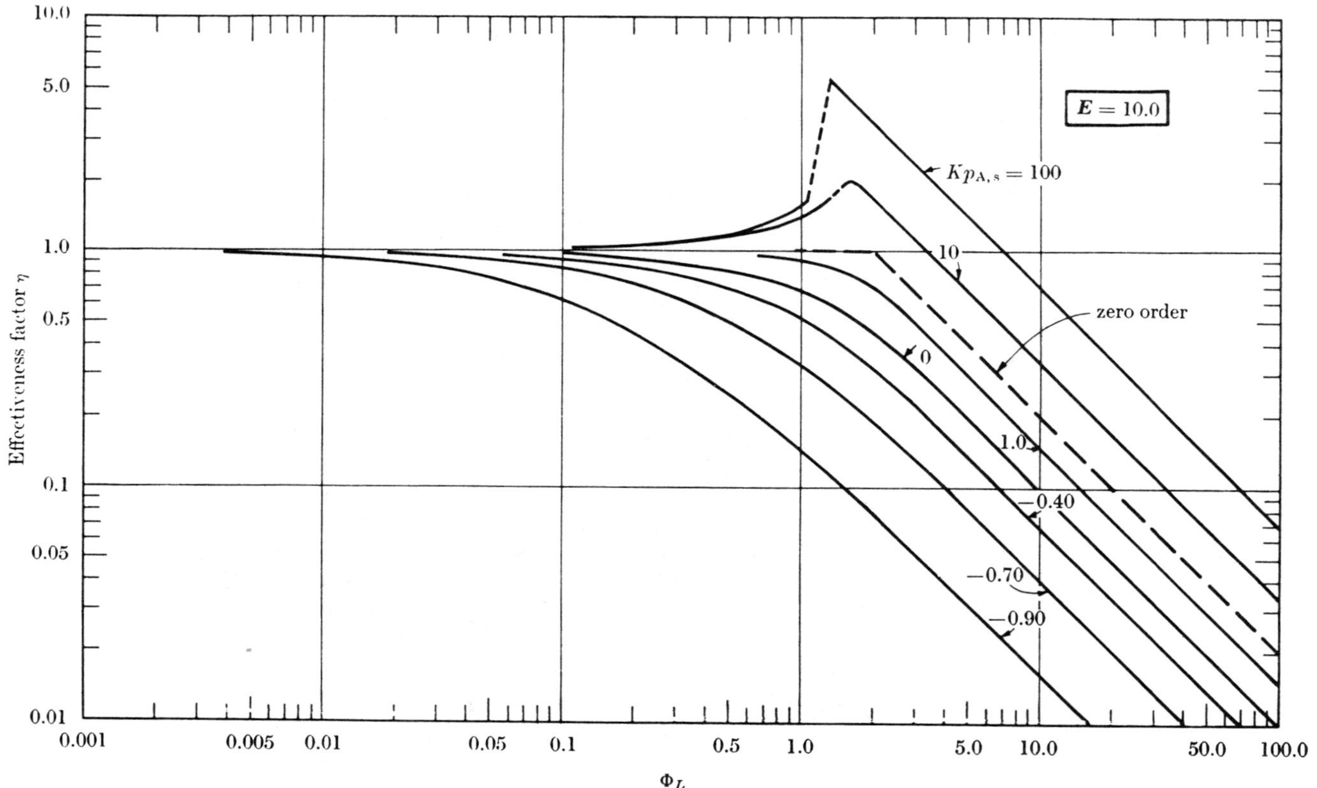

Figure 4.4. Effectiveness factor η as a function of Φ_L (Equation 3.32). $E = 10$ (Equation 4.8). Type II kinetics in flat plate geometry (Equation 4.5). Roberts and Satterfield [273].

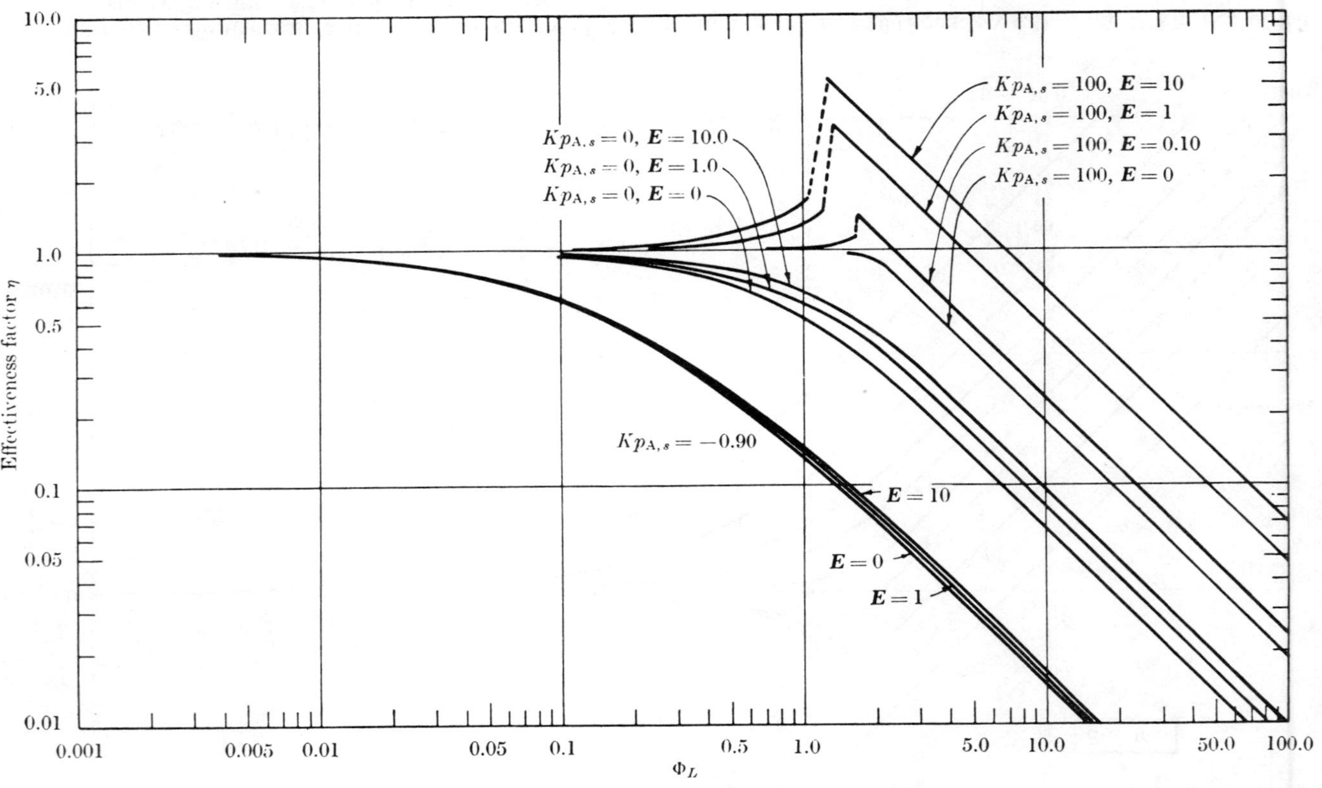

Figure 4.5. The η versus Φ_L relationship for representative combinations of $Kp_{A,s}$ (Equations 4.6, 4.7) and E (Equation 4.8). Φ_L is given by Equation 3.32. Type II kinetics in flat plate geometry (Equation 4.5). Roberts and Satterfield [273].

of $Kp_{A,s}$. For $Kp_{A,s} = -0.90$, the $\mathbf{E} = 0$ curve lies lowest, the curve for $\mathbf{E} = 10$ lies highest, with $\mathbf{E} = 1$ intermediate. However, the three curves lie so close together that they are almost indistinguishable. When $Kp_{A,s} = 0$, the order of the curve is the same; the spread is somewhat greater but still rather small. For $Kp_{A,s} = 100$, curves for $\mathbf{E} = 0$, 0.10, 1, and 10 are shown. The spread between these curves is large, but the order is the same as previously. The fact that the curves for high values of $\mathbf{E}$ lie above those for low values of $\mathbf{E}$ is consistent with the interpretation of $\mathbf{E}$ as a stoichiometric excess. When $\mathbf{E}$ is large, p_B is in excess throughout the pellet and the product $p_A p_B$ declines more slowly than it would if B were present in stoichiometric amount.

The region in which effectiveness factors greater than unity can theoretically occur in isothermal catalyst pellets is probably actually encountered in a number of hydrogenation reactions. Since hydrogen has a relatively high diffusivity and is usually present in excess over the stoichiometric concentration, values of $\mathbf{E}$ in the range of 10 to 100 are common. High values of $Kp_{A,s}$ imply relatively high adsorptivity and/or high partial pressure of the compound being hydrogenated. Adsorption coefficients for the species being hydrogenated that are high relative to hydrogen or to the products are fairly common — for example, that of ethylene in the hydrogenation of ethylene to ethane. However, hydrogenation reactions are exothermic and significant temperature gradients through the catalyst pellet can also cause effectiveness factors to exceed unity. Analyses of nonisothermal behavior combined with Langmuir-Hinshelwood kinetics are discussed in Section 4.2.2. This type of problem is complex, but separate analyses of either nonisothermal behavior with simple kinetics or complex kinetics under isothermal conditions may make it possible in many cases to judge which is the more probable contribution to unusual behavior.

Use of Figures 4.3, 4.4, and 4.5 is illustrated in Example 4.4. Some of the effects of initial reactant concentrations inside and outside the catalyst pellet upon the ultimate steady-state behavior were explored by Hartman [135]. It appears that only the initial condition of a high concentration of reactant B and a low concentration of (strongly adsorbed) reactant A inside the catalyst pellet will give values of the higher of the two steady-state effectiveness factors. This behavior is reasonable from the standpoint of the Langmuir-Hinshelwood model because component A in the presence of a large excess of B disappears by reaction close to the exposed surface of the catalyst and consequently the average concentration of A at steady state is low. Since A is strongly adsorbed, the reaction rate is greater than it would be at higher gas-phase concentrations of A at which B is in effect squeezed off the surface. This situation must occur in many hydrogenation reactions where it is common practice in starting up a reactor to contact the catalyst with hydrogen before admitting the other reactants. Hydrogen (component B) is usually the more weakly adsorbed reactant.

Two conditions that result in the lower values of the effectiveness factor are high initial concentration of both reactants or a low concentration of reactant B and a high concentration of A. Physically, the large concentration of A initially present plus its high adsorptivity allows very little B to be adsorbed on the catalyst, and thereby the reaction rate is retarded. This indicates the potential consequences of a sudden temperature drop in a system where reaction was proceeding at the higher rate in the multiple-valued region. With the consequent rate decrease, the effectiveness factor would decrease toward unity. The resulting increase in the partial pressure of A inside the pellet to approach that outside could force the reaction to the lower stable reaction rate when the system returns to the original temperature. Another initial condition that results in the lower effectiveness factor is a low initial concentration of both reactants inside the pellet. In practice, it is found that the steady-state hydrogenation activity of a catalyst may be profoundly affected by the startup conditions. For maximum activity and long life, the catalyst is usually brought in contact first with hydrogen and only subsequently with the material to be hydrogenated. The presence of hydrogen, either chemisorbed or within the interstices of the metal, seems to be necessary for good activity of a catalyst such as nickel; this is provided by this method of startup. This approach also helps minimize the formation of deactivating carbonaceous residues on the catalyst which are enhanced in the absence of hydrogen. Although such chemical factors undoubtedly are the cause of many of the observed effects of initial reaction conditions on steady-state activity, the present analyses demonstrate that physical factors may sometimes be the cause instead.

4.2.2 *Complex Kinetics. Nonisothermal Reaction*

These systems are complex to analyze and some of the assumptions made in order to obtain solutions may seriously restrict their applicability. Hutchings and Carberry [157] consider the kinetic expression for first-order irreversible reaction represented by Equation 4.4, in which the rate constant is a function of temperature but the adsorption coefficients are taken to be independent of temperature. Mitschka and Schneider [222] consider the same kinetics for the case in which k in

Equation 4.4 is represented as $k'K_A$. The adsorption constant for the reactant K_A is taken to be significant but adsorption constants for products or other species are assumed to be negligible.

The effect of temperature on both k' and K_A is represented by Arrhenius-type expressions. Since adsorption constants decrease with increasing temperature, one interesting conclusion is that for exothermic reactions the retardation effect of reactant adsorption is diminished by temperature gradients; e.g., for large positive values of $Kp_{A,s}$, the true effectiveness factor is higher than would be calculated assuming isothermal behavior. Analogous effects occur with endothermic reactions, but they are much less pronounced at low effectiveness factors.

When temperature gradients are substantial, the effect of geometry on performance can become significant; and the interchange of Φ_s and $9\Phi_L$ for calculation purposes becomes less accurate. The apparent (observed) activation energy may vary widely, from a value exceeding the true value to less than one-half the true value, as shown by the analyses of Schneider and Mitschka [307] and Ostergaard [238].

Example 4.4 Estimation of Effectiveness Factor, η; Isothermal, Complex Irreversible Reaction

Hougen and co-workers [338] studied the kinetics of the reaction

$$\underset{\text{(codimer)}}{i\text{-}C_8H_{16}} + H_2 \rightarrow i\text{-}C_8H_{18}$$

over a nickel-on-kieselguhr catalyst, between 200° and 325°C and 1 and 3.5 atm total pressure. Their data were best described by a rate equation of the form

$$r = k_2 p_H p_U/(1 + K_H p_H + K_U p_U + K_S p_S)^2.$$

In the above expression, H refers to hydrogen, U to codimer, and S to iso-octane. At 200°C, the following values were found: $K_H = 0.383$ atm^{-1}, $K_S = 0.489$ atm^{-1}, and $K_U = 0.580$ atm^{-1}.

Consider run 3d of this study, which was made at an average temperature of 200°C, and with the following partial pressures: $p_{H,s} = 2.45$ atm, $p_{U,s} = 0.53$ atm, and $p_{S,s} = 0.515$ atm. The observed rate of reaction was 0.0320 g-mol/(g catalyst)(hr). The catalyst was in the form of cylinders 0.292 cm in diameter and 0.269 cm long; the bulk density was 1.39 g/cm³.

In accordance with the rule for choosing component A, let A be U (codimer). For the purpose of calculating K, assume that the effective diffusivity is inversely proportional to the square root of the molecular weight. Then,

$$\frac{D_U}{D_H} = \sqrt{\frac{2}{112}} \cong 0.134; \quad \frac{D_U}{D_S} = \sqrt{\frac{114}{112}} = 1.01.$$

From Equation 4.7

$$\omega = 1 + 0.383[2.450 - 0.134 \times 0.530] + 0.489\,[0.515 + 1.01 \times 0.530] = 2.42.$$

The second term in the first bracket has a negative sign because $\nu_H = -1$. From Equation 4.6

$$K = [0.580 + 0.134 \times 0.383 - 1.01 \times 0.489]/2.42 = 0.137/2.42 = 0.0565 \text{ atm}^{-1}$$

$$Kp_{A,s} = 0.0565 \times 0.530 = 0.0300.$$

From Equation 4.8

$$E = [(2.450/0.134) - 0.530]/0.530 = 33.6.$$

The values of $Kp_{A,s}$ and E show that the effectiveness factor could not have been greater than unity. Further, since $Kp_{A,s}$ is essentially zero and the value of E is very large, the $\eta - \Phi_L$ relationship will be very close to that for a first-order reaction. The effectiveness factor can, therefore, be accurately estimated from the $Kp_{A,s} = 0$ line in Figure 4.4. Φ_L is estimated as follows: The quantity L will be approximated as the ratio of the volume to the external surface of the catalyst particles. This is about 0.0472 cm.

$$c_{A,s} = \frac{p_{A,s}}{RT} = 0.530/(82.06 \times 473) = 1.37 \times 10^{-5} \text{ g-mol/cm}^3,$$

$$-\frac{1}{V_c}\frac{dn}{dt} = (0.0320/3600) \times 1.39 = 1.24 \times 10^{-5} \text{ g-mol/(cm}^3 \text{ catalyst)(sec)}.$$

Estimation of D_{eff} is difficult since the physical characteristics of the catalyst — i.e., surface area, pore-size distribution, porosity, etc. — were not recorded. For the purpose of illustration we will assume diffusion by the Knudsen mode, assume that the porosity θ is about 0.50, and that the catalyst has a tortuosity τ of 2. Since kieselguhr is a fairly low-area catalyst support, we will take $r_e = 500$ Å.

$$D_{K,eff} = \frac{D_K\theta}{\tau} = \left(\frac{0.50}{2}\right) 97 \times 500 \times 10^{-8} \sqrt{\frac{473}{112}} = 0.0248 \text{ cm}^2\text{/sec}.$$

$$\Phi_L = \frac{(0.0472)^2 \times (1.24 \times 10^{-5})}{(1.37 \times 10^{-5}) \times (0.0248)} = 0.081.$$

The diffusional retardation is insignificant and η approaches unity, even if the estimate of the effective diffusivity is substantially in error. The use of charts of Φ_L instead of Φ_s has little effect on the result if reaction is essentially isothermal (Section 4.4). Having established that the reaction behaves essentially as a first-order reaction in codimer, one could use the line for $\beta = 0$ in Figure 3.4, 3.5, 3.6, or 3.7, taking $\Phi_s = 9\Phi_L$, and obtain essentially the same value of η.

4.3 Effects of Geometry

In Section 3.2 it was pointed out that, in any mathematical treatment of effectiveness factors, the assumption of flat plate (slab) geometry as the catalyst shape simplifies the mathematical analysis, even though most catalysts would be much more closely approximated as a sphere. Aris [11] showed that the functions of η versus the Thiele modulus ϕ for a first-order reaction in a sphere, in an infinite flat plate, and in a cylinder of infinite length lie very close together when the characteristic dimension is defined as the ratio of the volume to the outside surface through which reactant can diffuse. This analysis in terms of the modulus Φ has been extended by Knudsen [172] to the case of a single reactant A in which adsorption of products or reactant may be significant — i.e., a reaction obeying the rate equation given by Equation 4.4. Results are shown in Figure 4.6 for the flat plate and the sphere which represent the minimum and maximum values of volume-surface ratios that will be encountered in any of the usual catalyst shapes. The $\eta - \Phi$ functions for all other shapes would be expected to fall between the two curves. The modulus Φ_L is defined by Equation 3.32 for both the flat plate and the sphere, in each case L being the ratio of volume to surface through which reactants have access. Since this ratio equals $\frac{1}{3}R$ for a sphere, $\Phi_s = 9\Phi_L$. As expected, the η versus Φ_L functions become identical at very low and very high values of Φ_L. The functions for spherical and flat plate geometry lie closest together for negative values of $Kp_{A,s}$ (strong product inhibition or highest-order reactions) and show the greatest deviation for a zero-order reaction ($Kp_{A,s}$ approaching infinity). For a specified value of Φ_L, the calculated effectiveness factor η is always greater in flat plate geometry than in spherical geometry. The maximum deviation between the two values of η as a function of $Kp_{A,s}$ is shown below. These occur for values of Φ_L in the range of 0.1 to 2.

$Kp_{A,s}$	$\left(1 - \frac{\eta_{\text{sphere}}}{\eta_{\text{flat plate}}}\right) \times 100$
∞ (zero order)	34.0
+5.0	27.2
0.0 (first order)	22.0
−0.70	20.3
−0.90	17.0
−0.98	15.9

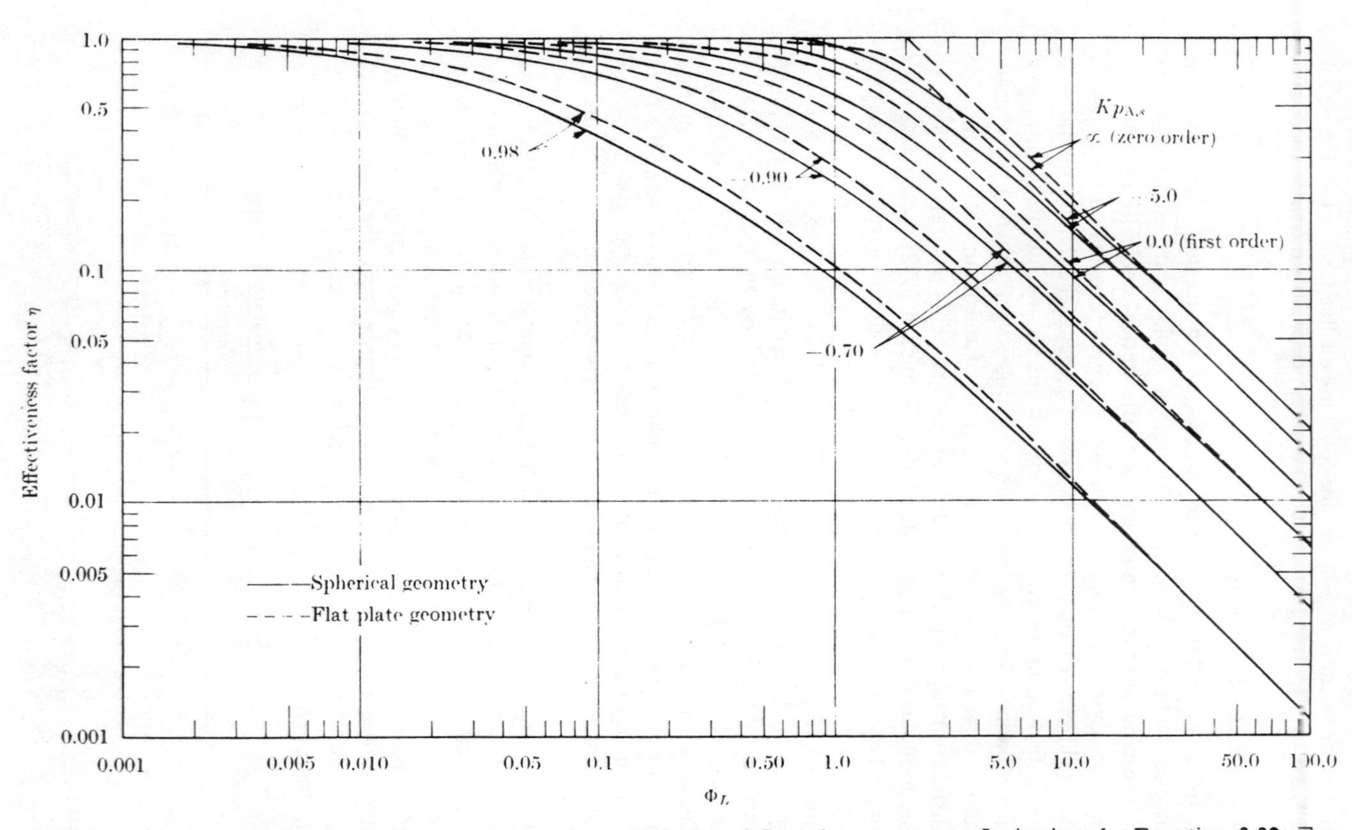

Figure 4.6. Comparison of effectiveness factors for spherical and flat plate geometry. Φ_L is given by Equation 3.32. For spheres $L = \frac{1}{3}R$. Knudsen, Roberts, and Satterfield [172].

For many calculations, the deviations are sufficiently small to be ignored.

To the extent that $\eta_{\text{flat plate}} \approx \eta_{\text{sphere}}$ for a specified value of either Φ_s or Φ_L, the moduli Φ_s and $9\Phi_L$ can be used interchangeably for calculations. For example, in Figures 3.4–3.7, for $\beta = 0$, the effectiveness factor corresponding to $\Phi_s = 90$ should be nearly equal to that for $Kp_{A,s} = 0$ (first-order kinetics) and for $\Phi_L = 10$ in Figures 4.2 and 4.6. The latter is identical to that for $\boldsymbol{C} = 0$ (irreversible reaction); $\boldsymbol{B} = 0$ (first-order) in Figure 4.9 at $\Phi_L = 10$ (see Section 4.4). In Example 4.3, Φ_L was found to be 0.108 and $Kp_{A,s} = -0.965$. Using Figure 4.2, which assumes flat plate geometry, $\eta_{\text{flat plate}}$ was estimated to be about 0.5. The actual shape used may be more closely approximated as a sphere. Figure 4.6 indicates that this value of η calculated for flat plate geometry was slightly higher than an analysis for spherical geometry would indicate, by a maximum of about 15 per cent.

Luss and Amundson [194] present a mathematical proof that for simple first-order isothermal reactions the spherical particle has the lowest effectiveness factor of all possible shapes having the same volume. They also present plots that in effect show how, for a given volume of catalyst particle, the effectiveness factor will vary as a cylindrical shape is made flatter or longer, or a rectangular parallelepiped is altered in dimensions. For example, at low effectiveness factors the rate of reaction on a cylindrical catalyst particle having an 8/1 ratio of length to diameter would be about 65 per cent greater than that on a spherical particle of the same volume. For maximum rate of reaction under diffusion-limiting conditions some shape other than spherical is in principle generally desirable, except perhaps for an exothermic reaction under conditions in which the effectiveness factor exceeds unity, when the spherical shape would act to maximize the temperature inside the catalyst particles. In practice, considerations of flow and pressure drop in catalyst beds may influence the choice of appropriate particle shape.

4.4 Reversible Reactions

Generalized analyses of effectiveness factors have usually assumed the reaction to be irreversible. Smith and Amundsen [325] made both theoretical and experimental studies of the hydrolysis of ethyl formate on Dowex-50, a cation exchange resin, in which the reverse reaction must be considered. For the simple first-order reversible reaction $A \rightleftharpoons X$, the same function of η versus ϕ_L is obtained as in the irreversible case if ϕ_L is defined as

$$\phi_L = L\sqrt{\frac{k_{\text{forward}}}{D_{\text{eff, reactant}}} + \frac{k_{\text{reverse}}}{D_{\text{eff, product}}}} \quad \text{(first-order, reversible).} \quad (4.9)$$

If D_{eff} is essentially the same for reactant and product, this becomes

$$\phi_L = L\sqrt{\frac{k_{\text{reactant}}(K_e + 1)}{K_e D_{\text{eff}}}} \quad \text{(first-order, reversible),} \quad (4.10)$$

where K_e is the equilibrium constant. Carberry [61] has analyzed the simple first-order reversible reaction for catalysts having a macro-micropore size distribution (Section 1.5.4). He points out that heat effects with reversible reactions will be less than those for the irreversible case because the effect of temperature on the equilibrium constant compensates in part for its effect on reaction rate. For example, in an exothermic reaction, the increased temperature toward the center of a pellet reduces the equilibrium constant while increasing the reaction-rate constants.

Kao [167] has developed a generalized graphical method utilizing the modulus Φ_L for calculating effectiveness factors in flat plate geometry for a simple reversible reaction that follows Langmuir-Hinshelwood kinetics. He also gives direction to earlier literature on treatments of reversible reactions. The reaction rate is assumed to obey the relationship

$$r = \frac{k[p_{\text{A}} - p_{\text{X}}(p_{\text{A},e}/p_{\text{X},e})]}{1 + K_{\text{A}}\, p_{\text{A}} + K_{\text{X}}\, p_{\text{X}} + \sum_i K_i\, p_i}. \quad (4.11)$$

Here A is the reactant, X the product, and the index i denotes any species present that can be adsorbed onto the surface of the catalyst but does not undergo reaction. The index e designates values at equilibrium. For reversible reactions, two new parameters are required over those used for irreversible reaction with a single reactant. One is $p_{\text{A},e}/p_{\text{A},s} = C$, the ratio of the equilibrium partial pressure of A to that at the outside surface of the catalyst. The other parameter is defined by

$$B = \frac{K(p_{\text{A},s} - p_{\text{A},e})}{1 + Kp_{\text{A},e}}, \quad (4.12)$$

where

$$K = [K_{\text{A}} - K_{\text{X}}(D_{\text{A}}/D_{\text{X}})]/\omega \quad (4.13)$$

and

$$\omega = 1 + K_{\text{X}}[p_{\text{X},s} + p_{\text{A},s}(D_{\text{A}}/D_{\text{X}})] + \sum_i K_i\, p_i. \quad (4.14)$$

The generalized definitions of K and ω as expressed by Equations 4.6 and 4.7 reduce to Equations 4.13 and 4.14, respectively, for a first-order reversible reaction. Note that their values are independent of whether or not the reverse reaction is considered. For an irreversible reaction, the parameter $\boldsymbol{B}$ in Equation 4.12 reduces to $Kp_{A,s}$, having the same definition as previously. Recall that an irreversible zero-order reaction corresponds to $Kp_{A,s}$ approaching infinity, a first-order reaction to $Kp_{A,s}$, approaching zero, and irreversible higher-order reaction to negative values of $Kp_{A,s}$. The minimum possible value is -1, corresponding to very strong product adsorption effects. The maximum and minimum values of $\boldsymbol{B}$ are, likewise, plus infinity and minus 1. The value $\boldsymbol{B} = 0$ corresponds to a simple first-order reversible reaction. The ratio of $P_{A,e}/P_{A,s} = \boldsymbol{C}$ will vary between 0 (for irreversible reaction) and 1 (for reaction at equilibrium and, therefore, corresponding to a zero rate). The absolute value of $\boldsymbol{B}$ is always less than that of $Kp_{A,s}$. Figures 4.7 and 4.8 present graphs of η versus Φ_L, for values of $\boldsymbol{C}$ of 0.5 and 0.9 and for a family of curves of $\boldsymbol{B}$. Figure 4.9 is a crossplot on $\eta - \Phi_L$ coordinates of families of curves of $\boldsymbol{C}$, for values of $\boldsymbol{B}$ of -0.98, 0, and 50. A graph of η versus Φ_L for $\boldsymbol{C}$ of 0.1 differs little from Figure 4.2. Graphs for values of $\boldsymbol{C}$ of 0.3 and 0.7 have been published [167]. The limiting case of $\boldsymbol{C} = 0$ represents irreversible reaction.

As has been pointed out by previous workers, neglecting to allow for reverse reaction results in the calculated effectiveness factor being higher than the true value. The two examples below illustrate this point and demonstrate the use of the graphs.

Example 4.5 Estimation of Effectiveness Factor η: Isothermal, Reversible Reaction

Gupta and Douglas [128] report studies on isobutylene hydration to t-butanol on Dowex 50W cation exchange resin. Water is present in large excess and the reaction rate is first order with respect to isobutylene concentration. The reverse reaction may be assumed to be first order with respect to t-butanol. They report that the equilibrium constant at 100°C corresponds to a minimum of 94 per cent conversion of isobutylene and they assumed reaction was essentially irreversible under their experimental conditions. For a simple first-order reaction, $Kp_{A,s} \to 0$, $\boldsymbol{B} \to 0$, $\boldsymbol{C} = p_{A,e}/p_{A,s} = 0.06/0.94 = 0.064$.

Calculate the effectiveness factor for a set of their experimental conditions corresponding to the greatest degree of diffusional limitations (highest temperature and largest particles). At 95°C on 48–65 mesh resin (radius in the swollen state = 0.0213 cm), the rate was about 55×10^{-3} g-mol/hr · g resin at a surface concentration of 0.0172×10^{-3} g-mol isobutylene/cm^3 of swollen

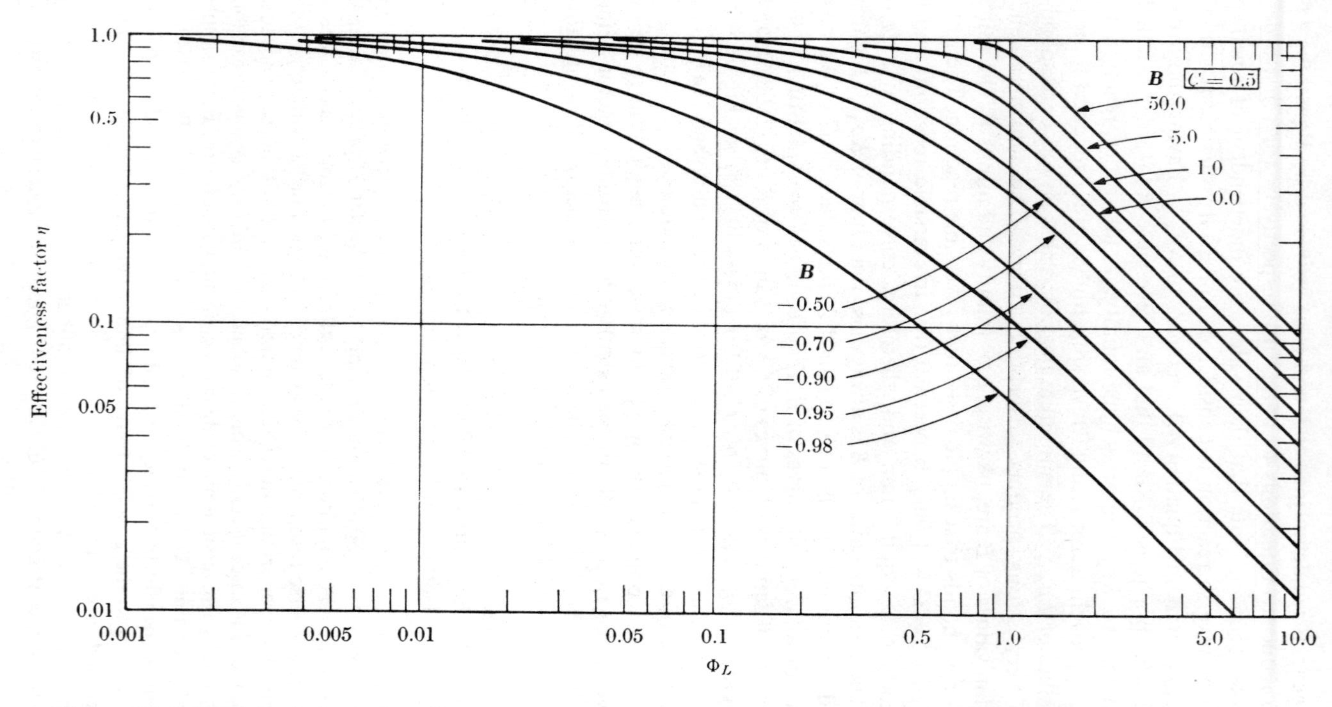

Figure 4.7. Effectiveness factor as a function of Φ_L (Equation 3.32) for reversible first-order reaction. $C = P_{A,e}/P_{A,s} = 0.5$. B defined by Equation 4.12. Kao and Satterfield [167].

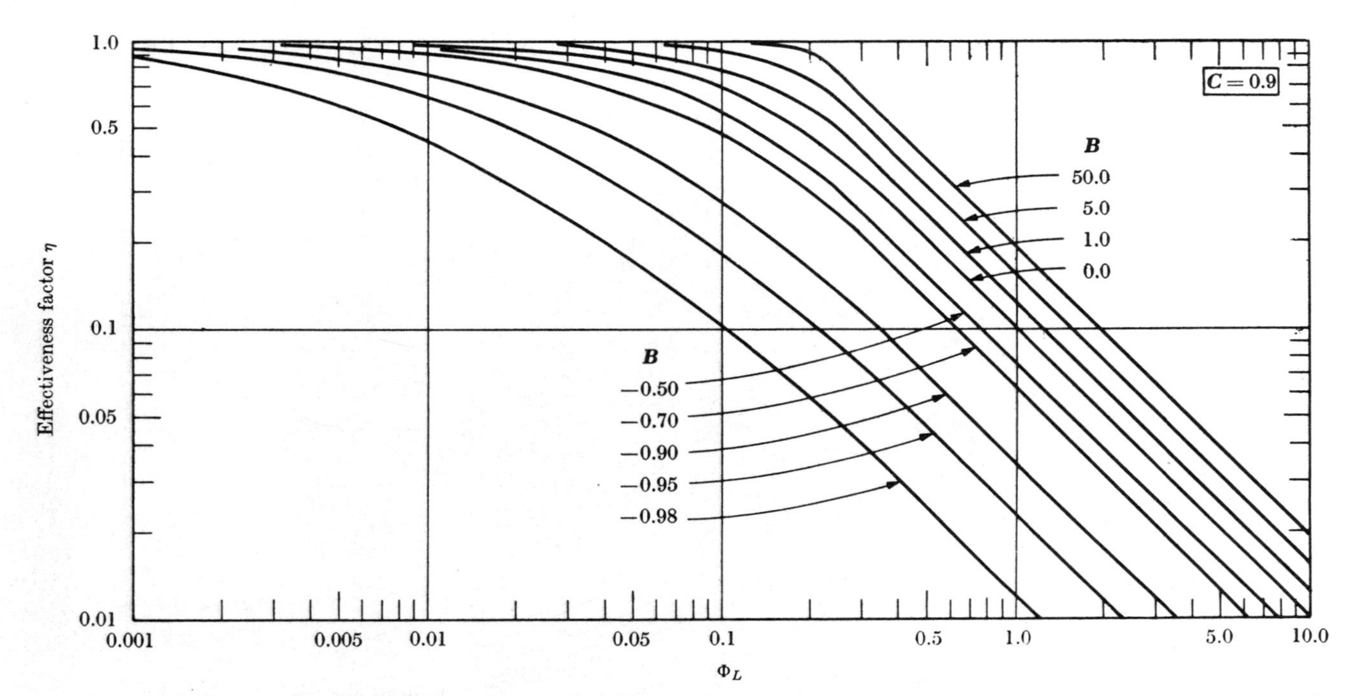

Figure 4.8. Effectiveness factor as a function of Φ_L (Equation 3.32) for reversible first-order reaction. $C = P_{A,e}/P_{A,s} = 0.9$. B defined by Equation 4.12. Kao and Satterfield [167].

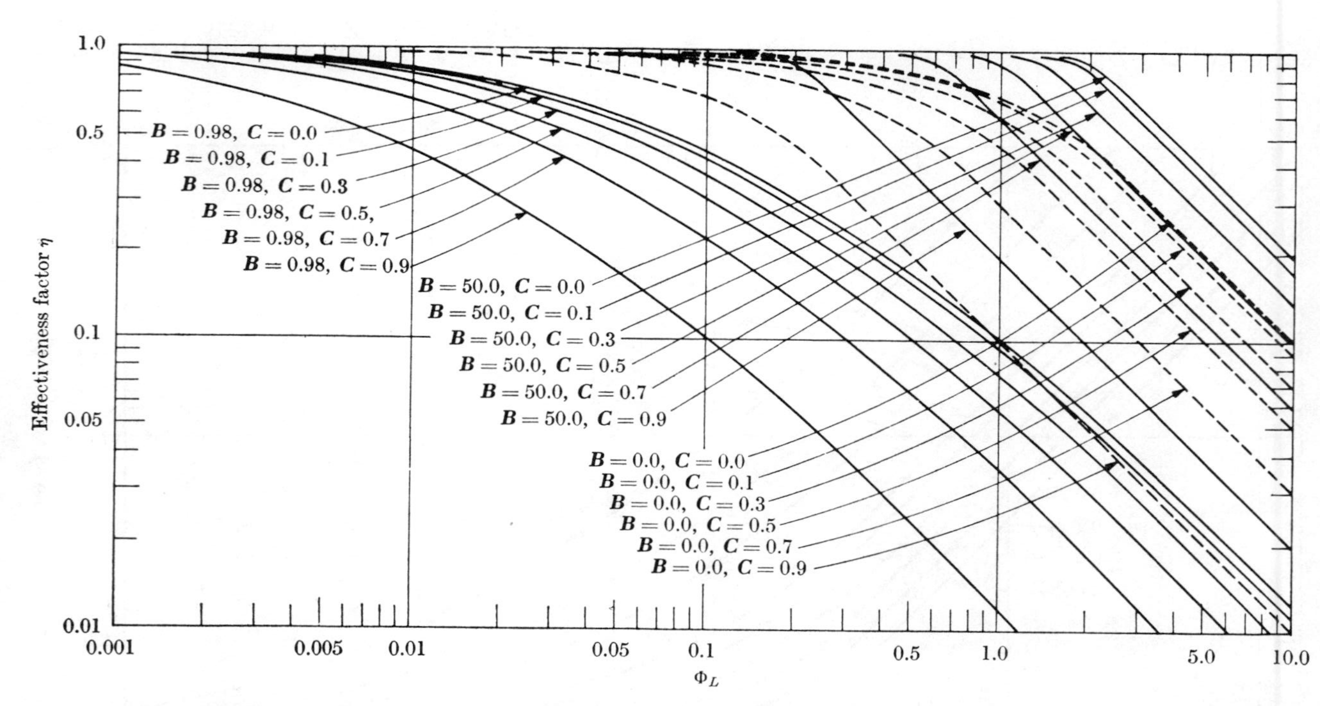

Figure 4.9. Effectiveness factor for reversible first-order reaction. The η versus Φ_L relationship for representative combinations of B and C. $C = P_{A,e}/P_{A,s}$. B defined by Equation 4.12. Kao and Satterfield [167].

resin. D_{eff} at 95°C was 1.6×10^{-5} cm/sec. The rate was specified per gram of anhydrous resin whereas we wish it on the basis of swollen resin volume. The degree of swelling was not specified, but for illustrative purposes we may assume about 1 cm³ of swollen resin per gram of anhydrous resin.

$$\Phi_s = \frac{R^2}{D_{eff}}\left(-\frac{1}{V_c}\frac{dn}{dt}\right)\frac{1}{c_s}$$

$$= \frac{(0.0213)^2}{1.6 \times 10^{-5}} \frac{(55 \times 10^{-3})}{3600} \frac{1}{1.72 \times 10^{-5}}$$

$$= 25.2.$$

To use the chart for flat-plate geometry, we take $\Phi_L = \frac{1}{9}\Phi_s = 2.8$. From Figure 4.9, interpolating between the curves for $\boldsymbol{C} = 0$ and $\boldsymbol{C} = 0.1$, and $\boldsymbol{B} = 0$, $\eta = 0.34$. Comparison of the curves shows that neglecting to allow for the reverse reaction results in a calculated value of η high by only about 5 per cent if equilibrium corresponds to 94 per cent conversion. Note that the degree of error incurred is independent of the value of η in this asymptotic region. In this case the reaction is taken to be simple first order in each direction, hence $\boldsymbol{B} = 0$ and it is unnecessary to know the effective diffusivity of the product. This, however, will be needed for more complex kinetic expressions.

Example 4.6 Estimation of Effectiveness Factor η: Isothermal, Reversible Reaction

Consider a case in which the combination of catalyst dimension, L, effective diffusivity, D_{eff}, observed reaction rate $(-1/V_c)(dn/dt)$, and reactant concentration at the outside surface, $p_{A,s}$, result in a value of $\Phi_L = 1.0$. We will suppose that the reaction is retarded by product adsorption to such a degree that $K = -0.7$, and we will take $p_{A,s} = 1$. If we assume that the reaction is irreversible, from Figure 4.2 the effectiveness factor η corresponding to $\Phi_L = 1.0$ and $Kp_{A,s}$ of -0.7 is 0.50.

However, we now discover that the reaction is in fact reversible and that $p_{A,e}$ is, let us say, 0.5. Then,

$$\boldsymbol{C} = p_{A,e}/p_{A,s} = +0.5$$

and

$$\boldsymbol{B} = \frac{K(p_{A,s} - p_{A,e})}{1 + Kp_{A,e}} = \frac{-0.7(1-0.5)}{1+(-0.7)(0.5)} = -0.54.$$

From Figure 4.7, the effectiveness factor is seen to be 0.35. Note that allowing for reverse reaction does not change the value of K, but changes the values of both $\boldsymbol{B}$ and $\boldsymbol{C}$. If $p_{A,e}$ were 0.9, then

$$\boldsymbol{C} = +0.9,$$

and

$$B = \frac{-0.7(1-0.9)}{1+(-0.7)(0.9)} = -0.19.$$

From Figure 4.8, $\eta = 0.09$ as opposed to 0.50 if reverse reaction were neglected. Calculations such as this indicate that when reaction is occurring near equilibrium (relatively large values of C) neglect of the reverse reaction can result in erroneously high effectiveness factors being calculated. A compensating factor, however, is that under such conditions the observed reaction rate will tend to be small, resulting in a low value of Φ_L.

4.5 Change in Volume on Reaction

The above mathematical developments assume that the diffusion flux can be represented by Fick's first law with a constant effective diffusivity independent of the nature of the reaction. In the case of bulk gas-phase diffusion, this is justified for equimolal counterdiffusion of two components, which occurs when there is no change in the number of moles on reaction, and is a suitable approximation for the non-equimolal case if one reactant or an inert gas is present in great excess. If mass transfer occurs by Knudsen diffusion only, then a change in the number of moles on reaction has no effect on the effective diffusivity as expressed by Fick's first law, but the total pressure may vary through the pellet, thus modifying the concentration gradient (see below). A change in number of moles on reaction causes a net molar flow either into or out of the porous catalyst. If diffusion occurs in the bulk or transition region, the pressure gradient through the pellet will usually be negligible, but an increase in number of moles makes it more difficult for reactant to diffuse into the catalyst and, hence, decreases the effectiveness factor. A decrease in number of moles has the opposite effect. At constant total pressure the magnitude of this effect for simple power-law kinetics and bulk diffusion depends only upon the stoichiometric coefficient of the reaction, ν ($A \rightarrow \nu B$), and upon the mole fraction of A in the fluid at the outside pellet surface, $Y_{A,s}$, so the influence of volume expansion or contraction can be expressed by one additional modulus. Thiele [333] and Hawthorne [136] computed some relationships, but the most general treatment is that presented recently by Weekman and Gorring [363]. They analyzed zero-, first-, and second-order reactions in spheres at *constant total pressure*, and expressed the influence of volume expansion on η as a function of a "volume change modulus" $\theta = (\nu - 1) Y_{A,s}$, in addition to ϕ_s. For each of the three kinetic expressions, they present plots of η versus ϕ_s for various values of θ, and η'/η versus θ for various values of ϕ_s. η'/η is the ratio of the effectiveness factor with volume change to that calculated ignoring volume change. A representative plot for first-order reactions is shown in Figure 4.10. The curve labeled $\phi_s = \infty$ is almost independent of the reaction order. For any finite value of ϕ_s, the ratio of η'/η approaches zero with increasing value of θ. η' itself approaches 1 as θ approaches -1 since the increasing inward flow of molecules causes the internal

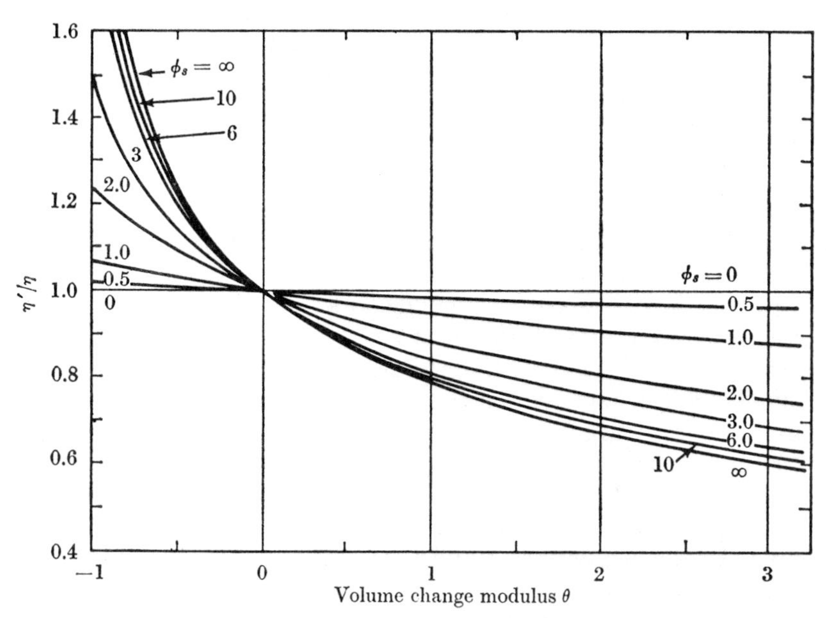

Figure 4.10. Effect of volume change on effectiveness factor. $\theta = (\nu - 1) Y_{A,s}$. First-order reaction, spherical geometry, bulk diffusion. Weekman and Gorring [363].

reactant concentration to approach that at the pellet surface. This shift in effectiveness factor caused by volume change is markedly affected by mole fraction of reactant. In a plug flow reactor the effect would thus be much more pronounced at the inlet than at the exit and the effect becomes negligible for a reactant highly diluted by nonreacting gases. The combined effect of volume change and nonisothermal behavior was also analyzed by Weekman [362].

Otani, Wakao, and Smith [241, 243] and, earlier, Wheeler [383], analyzed some of the effects on intraparticle pressure gradients caused by a change in number of moles on reaction for *isothermal, first-order, irreversible reaction.* They consider the range of diffusion from Knudsen to the bulk mode using the single-pore model of a catalyst pellet. For Knudsen diffusion, the pressure difference between outside and center of the pellet is given by [241]

$$\frac{P_c - P_s}{P_s} = Y_{A,s}(\nu^{1/2} - 1)\left(1 - \frac{1}{\cosh \phi_L}\right), \tag{4.15}$$

where $\phi_L = L\sqrt{2k_s/r_e D_K}$ (see Section 3.2). The pressure difference is maximized in the Knudsen diffusion regime and at low effectiveness factors. The maximum pressure gradient possible, at high values of ϕ_L, is given by the surprisingly simple relationship

$$\frac{P_c - P_s}{P_s} = Y_{A,s}(\nu^{1/2} - 1). \tag{4.16}$$

Here P_c is the pressure at the pellet center and P_s that at the outside surface.

For first-order irreversible reaction and Knudsen diffusion, a change in total pressure through the pellet has no effect on the effectiveness factor; this will not necessarily be true for other types of kinetics. Otani *et al.* [243] analyze some of the effects of ν and ϕ_L on pressure drop and on the effectiveness factor through the transition range. Pressure increase in the interior of a catalyst may be sufficient in a few cases to cause significant physical degradation, particularly under unsteady-state or cycling conditions. This seemed to be the cause of rapid breakup of a particular rather soft industrial catalyst used under cycling conditions where a liquid reactant was converted to a vapor product by exothermic reaction.

4.6 Other Treatments of Effectiveness Factor, Isothermal Conditions

4.6.1 *Asymptotic Solutions*

At sufficiently large values of the modulus ϕ or Φ corresponding to reactant concentrations in the center of the catalyst particle approaching zero, the effectiveness factor becomes inversely proportional to the modulus and mathematical expressions relating the effectiveness factor to the modulus become considerably simplified. These are frequently termed "asymptotic solutions" or "asymptotic forms," and have been used in a number of ways. One is to modify the Thiele modulus by, in effect, incorporating the intrinsic kinetic expression into it in such a way as to cause all the asymptotes to superimpose, thereby bringing the families of curves in such figures as Figure 3.2 and Figures 4.2–4.5 closer together, and replacing them by a single curve at high values of ϕ or Φ. This seeming simplification, however, is purchased at the cost of complicating the modulus, and it becomes difficult to interpret the modulus in terms of physical reality. Attempts have also been made to define criteria for absence of diffusion effects by extrapolation of various asymptotic solutions to a value of $\eta = 1$. This is exact for the isothermal zero-order reaction, but becomes increasingly unreliable as one considers higher-order reactions, those in which product molecules are highly adsorbed, nonisothermal situations, etc. These approaches are useful when one is certain he is dealing with low effectiveness factors, but they do not indicate when deviations from the asymptotic model become significant. Such approaches are discussed in some detail in a book by Petersen [251] and in a number of papers [12, 13, 33, 34, 61, 250].

4.6.2 *Macro-Micropore Systems*

If a catalyst pellet is prepared by compressing a fine powder which itself has a fine porous structure, the resulting pellet consists of a network of macropores comprising the residual space between the original powder particles, from which micropores branch out, the lengths of the micropores being approximately that of the diameter of the powder particles. This is probably a good representation of relatively high-porosity pellets in which the pressing operation has not sealed off the micropores and which have not been subjected to chemical, heat, or other treatments that would alter the structure (Section 1.5).

Mingle and Smith [219] and Harriott [134] developed a mathematical model for such "macro-micropore" systems which views the diffusion as occurring in a main macropore from which micropores branch out along the wall. Under ambient conditions diffusion in the micropores is usually of the Knudsen type, while that in the macropores is predominantly in the bulk or transition range. Mingle and Smith calculate microeffectiveness factors η_{micro} for several distribution functions for the micropore radii and interpret the effect upon the macroeffectiveness factor η for irreversible reaction and isothermal as well as nonisothermal conditions. Using the same model, Carberry [61] treats the macro-micro effectiveness factor for the simple reversible isothermal case with allowance for a mass transfer limitation to the outside of the large pellet. In this model, the over-all effectiveness factor is equal to the product of the effectiveness factor for the macrostructure and that for the microstructure, i.e., $\eta_{\text{macro}} \times \eta_{\text{micro}}$.

However, if a fine powder is used in compacting the catalyst pellet, the lengths of the micropores will consequently be short, and η_{micro} would generally be expected to be essentially unity. This may be readily shown by a simple calculation. Most industrial reaction rates on pelleted catalysts are of the order of about 10^{-6} g-mol/sec·cm^3 of pellet. At rates above about 10^{-5} g-mol/sec·cm^3 of pellet (2.25 lb-mol/hr·ft^3), severe mass and heat transfer problems will usually be encountered with pellets of typical size. If the modulus Φ_s is about 0.5 or less, η will be nearly unity unless one is dealing with a reaction highly inhibited by product adsorption. D_{eff} for a porous catalyst will vary with pressure, temperature, and nature of diffusing species, but usually will not be less than 10^{-3} cm^2/sec. The concentration of a single pure reactant at 1 atm pressure and 400°C is 1.8×10^{-5} g-mol/cm^3.

From the definition of Φ_s,

$$R = \left[\frac{\Phi_s D_{eff}\, c_s}{\left(\dfrac{-1}{V_c}\dfrac{dn}{dt}\right)}\right]^{1/2} = \left[\frac{(0.5)(10^{-3})(1.8 \times 10^{-5})}{10^{-5}}\right]^{1/2}$$

$$= 0.03 \text{ cm}.$$

Even for these relatively extreme conditions, η_{micro} will be essentially unity unless the original particles are substantially greater than 0.06 cm diameter. A similar calculation can be made for any other set of circumstances. Calculations of this sort also show that diffusion limitations are not significant in most fluidized-bed reactors. One case in which η_{micro} may be much less than unity exists when the micropores are of a size similar to the dimensions of reactant or product molecules, as in certain molecular sieve catalysts, in which case the effective diffusivity with respect to certain species is extremely low.

Altering the porous catalyst structure so as to increase the effective diffusivity might in some cases make possible an increase in the capacity of a packed-bed reactor. For example, a homogeneous xerogel catalyst, being operated at an effectiveness factor substantially less than unity, could be replaced by a bimodal pore-size structure that might provide substantially greater reaction rate per unit volume.

Cunningham and Smith [82] analyze the case of a series of catalyst pellets prepared in the laboratory by compression of a catalyst powder to different degrees of compaction. Generally, the fine powders are found not to be crushed, and the micropore structure is essentially unaltered by variations in degree of compression. Since these micropores comprise typically 90 per cent or more of the total area, the surface area for reaction is nearly proportional to the unit mass. Hence, increasing the pellet density by this process increases the total surface area per unit volume of reactor, but decreases the effective diffusivity by decreasing macropore size. The maximum rate of reaction *per unit volume* of reactor would thus occur at a pellet density that may correspond to an effectiveness factor substantially less than unity. It is uncertain, however, whether the micropores would remain unaltered under the conditions typically used in industrial manufacture of catalysts. Binders and other ingredients may cause plugging and calcination may alter their structure. Basmadjian [22] reports optimum radius ratios for catalysts in the form of hollow cylinders as a function of the Thiele modulus.

4.6.3 *Other Treatments of Effectiveness Factor*

In multicomponent systems in which diffusion occurs in the transition or bulk mode, representation of the diffusion process and estimation of the effectiveness factor can become very complex. Some treatment of this problem is given by Butt [50]. Krasuk and Smith [177] present a method of estimating the effectiveness factor if surface diffusion is significant. A more detailed analysis is presented by Foster and Butt [108], who consider nonisothermal as well as isothermal systems. A similar analysis is given by Mitschka and Schneider [221]. The melting of a bed of ice crystals by direct contact with low-pressure water vapor, involving heat transfer through a film of water on the ice, is analogous to the effectiveness factor treatment for first-order reaction in a porous catalyst. Brian, Smith, and Petri [47] have analyzed this process, important in desalination, by the heat transfer equivalent of the Thiele approach, and have shown the significant consequences for design. Coughlin [77] presents the extensive analogy between the Thiele treatment and heat transfer from a longitudinal fin on a heated tube.

4.7 Criteria for Insignificant Diffusion Effects

When reaction occurs in a porous structure, *some* concentration gradient will always exist, and the effectiveness factor for isothermal reaction of a single reactant will never equal unity, except as a mathematically limiting case. A general criterion for absence of significant diffusion effects can be taken as those conditions for which η exceeds about 0.95 [372], but the accuracy with which this can be expressed in terms of the parameter Φ_s or Φ_L depends largely upon one's degree of knowledge concerning the intrinsic kinetics, the accuracy with which the effective diffusivity can be estimated, and the degree to which the catalyst is isothermal. There is no substitute for calculating the value of η or limits on the probable values of η for any specific case of interest by, for example, the various generalized charts available. Certain comments, however, may be helpful. For a single reactant in spherical geometry, the values of $\eta > 0.95$ for a second-order reaction correspond approximately to $\Phi_s \leq 0.3$; for first order, to $\Phi_s \leq 1.0$; and for zero order, to $\Phi_s \leq 6$. If $\Phi_s > 6$, diffusion effects will definitely be present, even for the unusual kinetic case of zero-order reaction; if $\Phi_s < 0.3$, diffusion effects will be insignificant unless the reaction rate is strongly inhibited by product. In flat plate geometry, values of $\eta \geq 0.95$ correspond to Φ_L values of ≤ 0.075 for second-order kinetics and ≤ 2.1 for zero-order kinetics. A possibly extreme case, which illustrates the importance of knowing the intrinsic kinetics, is the reaction of carbon dioxide with porous carbon at temperatures of about 1000°C, discussed in illustrative Example 4.3. The reaction is highly inhibited by the product carbon monoxide at this temperature and a value of $\eta = 0.95$ corresponds to a value of Φ_L of less than 0.01.

The above applies to isothermal conditions. For the first-order irreversible nonisothermal case, Weisz and Hicks [376] estimate that the criterion

$$1 > \Phi_s e^{[\gamma\beta/(1+\beta)]} \tag{4.17}$$

defines the conditions under which a reaction is free of significant mass and heat diffusion effects.

Example 4.7 Application of Criteria for Insignificant Diffusion Effects

Weisz [372] reports measurements of the rate of burnoff in air of carbonaceous deposits formed on a silica-alumina cracking catalyst. Oxygen consumption rates are reported for two particle sizes, 0.20-cm beads and 0.01-cm powder, at temperatures of 460°C and higher. At 460°C, the rate per unit mass of catalyst was the same for both particle sizes, demonstrating that the effectiveness factor for the 0.20-cm beads at 460°C was nearly unity. Significant differences between the rates on the two particle sizes began to appear at a temperature of about 475°C. The corresponding rate of oxygen consumption was about 4×10^{-7} g-mol/sec · cm³. The diffusivity of catalyst samples, using hydrogen and nitrogen on opposing faces, was measured and found to be 6.2×10^{-3} cm²/sec for hydrogen at 20°C. Prater [256] has shown that temperature gradients within the pellet are unimportant in this reaction. Compare these experimental results to a theoretical prediction of the rate of reaction at which diffusion effects would begin to become appreciable ($\eta \leq 0.95$).

Carbon combustion is about first order in oxygen and, in the presence of excess air, inhibition by products is probably negligible. The critical value of the parameter Φ_s will be about 1.0. Diffusion is by the Knudsen mode.

$$D_{\text{eff}} = 6.2 \times 10^{-3} \sqrt{\frac{748}{293} \times \frac{2}{32}} = 2.48 \times 10^{-3}\ \text{cm}^2/\text{sec}.$$

$$c_s = \frac{0.21}{22400} \times \frac{273}{748} = 3.42 \times 10^{-6}\ \text{g-mol/cm}^3.$$

Calculate the value of the maximum reaction rate in the 0.20-cm particles above which diffusion will become appreciable.

$$\frac{R^2}{D_{\text{eff}}}\left(-\frac{1}{V_c}\frac{dn}{dt}\right)\frac{1}{c_s} = \Phi_s \cong 1,$$

$$\frac{(0.20)^2}{2.48 \times 10^{-3}} \times \frac{1}{(3.42 \times 10^{-6}} \times \left(-\frac{1}{V_c}\frac{dn}{dt}\right) \cong 1,$$

$$\left(-\frac{1}{V_c}\frac{dn}{dt}\right) \cong 2.12 \times 10^{-7}\ \text{g-mol/sec} \cdot \text{cm}^3.$$

This compares closely with the value of about 4×10^{-7} found experimentally.

5 Poisoning, Reaction Selectivity, Gasification of Coke Deposits

5.1 Catalyst Poisoning

A poison is a species present in the feed stream or formed by reaction which is adsorbed onto the catalyst and inactivates it. Two opposite limiting cases of the behavior of a poison on an individual catalyst pellet can be visualized. In one the poison is uniformly distributed at all times throughout the pellet. This would occur if the poison were only weakly adsorbed onto the catalyst and were present in the feed stream in sufficiently large concentration that a steady-state surface concentration is quickly attained, or if it were formed by a reaction occurring with an effectiveness factor near unity, e.g., coke formation on a cracking catalyst when $\eta \rightarrow 1.0$. The opposite extreme occurs when the poison is present in the feed stream in small concentration and is so strongly adsorbed that the outside pore structure of a catalyst pellet becomes completely poisoned before the interior loses activity. This is the so-called "pore-mouth" poisoning. Since catalysts strongly adsorb many impurities, this behavior is common.

Wheeler [382, 383] mathematically analyzed these two limiting cases for isothermal first-order reaction in a cylindrical pore. Some of his results are shown in Figures 5.1 and 5.2. The fraction of the total reaction sites poisoned is represented by α, In the case of uniform poisoning, the intrinsic activity of the catalyst surface is assumed to be $k_s(1 - \alpha)$ throughout. For pore-mouth poisoning, it is assumed that for a pore of length L, the length αL nearest the mouth is completely poisoned and the remaining length $(1-\alpha)L$ retains the initial intrinsic activity.

If the effectiveness factor is essentially unity, in the "pore mouth" model the fraction of catalytic activity left is directly proportional to the fraction of surface unpoisoned, regardless of the distribution of the poison. This is illustrated by the diagonal line *A* in Figure 5.1. With low effectiveness factors, however, the two manners of distribution of poison

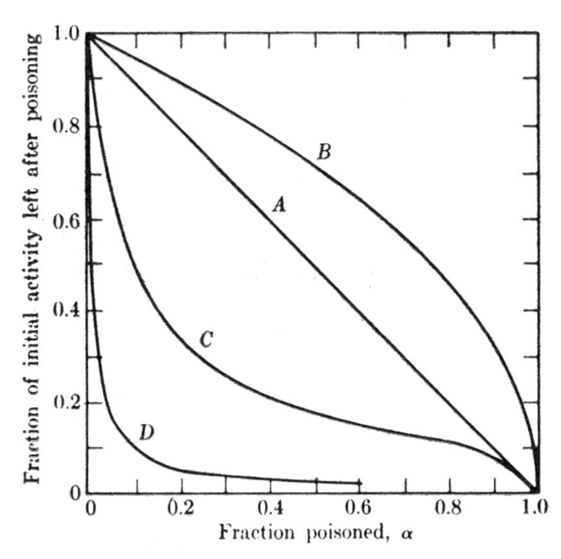

Figure 5.1. The ratio of the activity of a catalyst after poisoning to the activity before poisoning as function of the fraction of the surface poisoned. Wheeler [382, 383]. Curve A: $\eta \rightarrow 1.0$, any distribution of poison. Curve B: $\eta \rightarrow 0$, uniform distribution of poison. Curve C: $\eta = 0.1$, pore-mouth poisoning. Curve D: $\eta = 0.01$, pore-mouth poisoning. (Effectiveness factors refer to conditions before poisoning.)

affect the reduction of catalyst activity in opposite ways. If the poison is distributed uniformly, activity falls less than linearly with poison accumulation. A physical explanation is that a fraction of the poison is adsorbed in the center portions of the pellet which make very little contribution to the total reaction at low effectiveness factors. With pore-mouth poisoning, at low effectiveness factors the catalyst activity will drop very rapidly as the fraction of the surface poisoned increases. Physically, the poisoning of the pore mouths causes the reactants to diffuse deeper into the pellet before reaction can occur. Under extreme conditions (e.g., $\eta < 0.05$ and $\alpha > 0.2$), the rate becomes essentially the rate of diffusion through poisoned pore mouths under the maximum available concentration gradient, and the reaction shows the characteristics of a diffusion-limited process, e.g., activation energy corresponding to the effect of temperature on the diffusion coefficient, which typically corresponds to an apparent activation energy of about 1 to 3 kcal/g-mol for gases and 2 to 4.5 for liquids (see Chapters 1 and 2).

Figure 5.2 shows several cases calculated by Wheeler for pore-mouth

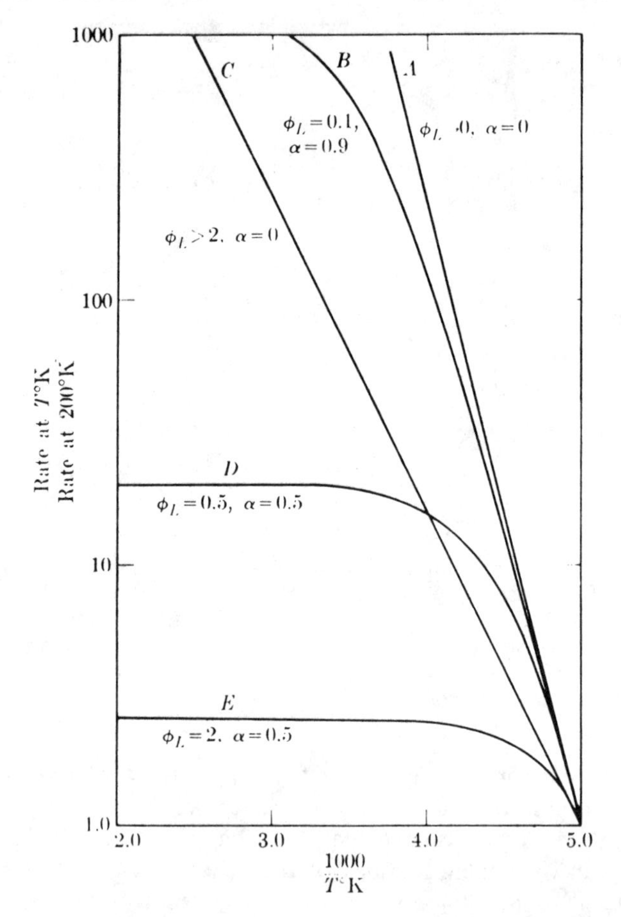

Figure 5.2. Predicted reaction rate (relative to rate at 200°K) versus $1/T$ for several combinations of Thiele modulus ϕ_L and fractional covering of the surface with poison, α (the poison is assumed to collect at the pore mouths). Wheeler [382, 383]. Note that ϕ_L is the value at 200°K; ϕ_L will increase rapidly with increasing temperature.

poisoning to show how the observed rate would be expected to change with temperature for different combinations of the Thiele modulus and fraction of surface poisoned. A reaction having 11 kcal/g-mol intrinsic activation energy is illustrated, and the rate of reaction relative to that at 200°K is calculated. Note that the values of ϕ_L on the curves apply at 200°K. As the temperature increases, ϕ_L will rapidly increase and the effectiveness factor drop. Curves *D* and *E* show the decrease in activation energy as temperature is increased, caused by the limitation of pore-mouth diffusion. Such a decrease will also be found, however, if mass transfer to the outside of the pellet surface becomes controlling. A few catalytic reactions also show a maximum in the intrinsic rate as temperature is increased. This occurs, for example, with hydrogenation of many olefins and also with the decomposition of hydrogen peroxide vapor on platinum [301]. This latter possibility can be analyzed by calculating the rate of mass transfer to the outside surface by methods described in Chapter 2.

The above analyses assume that the amount or distribution of poisons does not vary with time. In actual operation, however, a poison usually accumulates onto the active sites of the catalyst, causing a gradual drop-off in activity. It is clear that the observed reduction in reaction rate with time may be markedly affected by the manner in which the poison becomes distributed throughout the catalyst pellets. Carberry and Gorring [63] present an analysis showing how the fraction of a spherical catalyst pellet that is poisoned varies with time depending upon the extent to which the distribution of poison approaches the pore-mouth model ("shell-type poisoning") or the well-distributed model. They also include the effect of various degrees of mass transfer limitation to the outside of the catalyst pellet. The mathematical relationships are the same as those encountered in the gasification of carbon and are discussed in Section 5.3.2.

For pore-mouth type poisoning and no bulk mass transfer limitations, the fractional degree of poisoning in a sphere up to about 40 per cent or so is proportional to the square root of time on stream and observation of this type of relationship has sometimes been taken to indicate shell-type diffusion control. Carberry and Gorring show that other combinations of poison distribution and bulk phase mass transfer limitations can also result in essentially this type of relationship over wide degrees of poisoning, so this type of behavior is not of itself proof that shell-type poisoning is occurring. Nevertheless, the converse appears to be true. Van Zoonen [345] studied hydroisomerization of olefins on a silica-alumina-nickel sulfide catalyst. Microphotographs of catalyst cross sections showed the coke to have accumulated essentially in a peripheral shell and the rate of coke formation was inversely proportional to the quantity already present, which is equivalent to a square root proportionally with process time.

If poison indeed deposits preferentially in a thin layer near the outside of the particles, as is frequently the case, theory indicates that the apparent resistance to poisoning will be increased with decreased particle size; i.e., the observed rate of reaction will drop off more slowly with time. This in fact was found by Anderson, Karn, and Shultz [7] in a study of sulfur poisoning of a Fischer-Tropsch catalyst. The study of Van Zoonen showed little effect of temperature on the rate of reaction and an increase in average reaction rate with decreased pellet size or with increased pore volume of catalyst — all of which are in agreement with theory. Coke contents varied up to about 4 wt %. With a fixed bed of catalyst, the circumstances that lead to pore-mouth type poisoning should also cause the poison to accumulate preferentially in the pellets toward the bed inlet and the study of Van Zoonen showed a linear drop in coke content with distance through the bed; i.e., a diffuse boundary between coked and noncoked catalyst pellets moved downstream with time, simultaneous with the growth in each pellet of a coke deposit moving gradually toward the center.

The behavior of an isothermal fixed-bed reactor with shell-type poisoning is analyzed by Olson [233]. The extension of this model to an adiabatic fixed bed, as is encountered in regeneration of coked catalysts by combustion, is analyzed by Olson, Luss, and Amundson [234]. Murakami and co-workers [223], studied, both theoretically and experimentally, the parallel- and series-type reactions (Equations 5.10 and 5.1, respectively) in which in either case the reactant A is converted to desired product B or undesired coke C. The disproportionation of toluene on an alumina-boria catalyst to form xylene and benzene or coke and benzene was chosen as representative of parallel-type kinetics and the dehydrogenation of an alcohol on an alkaline-alumina catalyst to form an aldehyde which then formed coke, as representative of series-type reaction. One interesting result from both the experimental and analytical studies was the observation that with series-type reaction, under a moderate degree of diffusion limitation, coke deposition occurs preferentially in the interior of the pellet; with a high degree of diffusion limitation the coke deposition appears preferentially at the outside with either type of kinetics. Masamune and Smith theoretically analyze the effects of poisoning by the two same kinds of reactions as well as that produced by one of two independent reactions

(Equation 5.6) for isothermal [201] and nonisothermal [288] conditions, and these results are extended by Chu [71]. Sada and Wen [287] analyze the effect of the degree of catalyst poisoning of the pore-mouth type on the selectivity expected to be observed with the three types of complex reactions, for spherical, infinite flat plate, and infinite cylinder geometry, and for cases in which poisoning affects only one of two simultaneously occurring reactions, or both. Note that the combustion of carbon in an initially uniformly coked catalyst (Section 5.3) follows the same type of mathematical relationships as the deposition of poison onto a porous catalyst by a first-order process.

The theoretical analyses of poisoning usually assume that the accumulation of poison has no effect on the diffusion characteristics of the catalyst. However, these effects can be substantial with large coke contents. Suga and co-workers [330] studied the decrease in activity of 20 per cent chromia–80 per cent alumina cylindrical catalyst pellets (surface area unspecified) as caused by the dehydrogenation of *n*-butane at 555°C and 1 atm. From studies on powdered catalyst, the effectiveness factor on catalyst pellets 4.85×4.45 and 2.85×3.20 (diameter × length, mm) was found to be 0.74 and 0.85. From this and other information, the carbon deposit was thought to be essentially uniform through the pellets. Measurements of the effective diffusion coefficient as calculated by Fick's law, using nitrogen at 20°C and 1 atm, showed a decrease with increasing amounts of coke present, from 0.0195 cm^2/sec initially to 0.0152 with 7.3 wt % coke, to 0.0107 cm^2/sec with 14.4 wt % coke. Studies of a silica-alumina catalyst coked under nondiffusion limiting conditions showed no effect on diffusivity of quantity of coke in contents up to 1 wt % [244].

Several factors determine whether the coke deposit will be of the shell type, uniformly distributed, or something in between. (See Section 5.3.2.) A necessary, although not sufficient, condition for uniform distribution is the absence of diffusion limitations with respect to the reactant. This is relatively easy to establish in laboratory work with a single reactant; industrial processing, however, frequently involves mixtures of a wide variety of types of hydrocarbons, which vary greatly in their tendency to form coke during dehydrogenation, cracking, and similar reactions. This tendency is greatest with unsaturated hydrocarbons and least for paraffins. Thus, the coke-forming reaction might be diffusion limited with respect to certain hydrocarbons but not to others. Heavy metals catalyze coke-forming reactions, so shell-type deposits are readily formed on catalysts such as the nickel sulfide studied by Van Zoonen.

Crude petroleum contains varying amounts of metal porphyrins, especially of nickel and vanadium, which are found preferentially in the residual fraction from distillation. Concentrations are, for example, 73 ppm V and 25 ppm Ni in an Iranian topped residue [143]. In hydrodesulfurization of residual fractions, the nickel compounds are relatively weakly adsorbed onto the catalyst; nickel passes through the reactor relatively soon after commencing continuous operation, and nickel deposits are found to be distributed rather uniformly throughout the catalyst pellets. Vanadium is removed from the feedstock for much larger times onstream and is found preferentially in the outer layers of the catalyst pellets. These poisons reduce catalyst activity at least in part by blocking pores and reducing the effective diffusivity [143]. Upon regeneration, vanadium compounds may also flux with portions of the catalyst and thereby inactivate it. With a catalyst containing 7.2 wt % V_2O_5 from deposition it was found necessary in one case to increase reaction temperature after regeneration by 35°F above that used with fresh catalyst in order to obtain the same degree of reaction. With 12.6 wt % V_2O_5 on the catalyst, a 75°F temperature rise was required after regeneration [10].

5.2 Selectivity in Porous Catalysts

The discussion in Chapters 3 and 4 is concerned with a simple reaction that can be characterized as

$$A + B \longrightarrow \text{products.}$$

In practice, however, one may frequently be concerned with a complex of parallel and/or series reactions in which it is desired to maximize the concentration in the product of some intermediate. Several common cases found in practice are fairly well represented as a simple series of first-order reactions.

$$A \xrightarrow{k_1} B \xrightarrow{k_2} C \qquad \text{(Type III selectivity),} \tag{5.1}$$

where the intermediate B is the desired product. This is termed "Type III selectivity" by Wheeler. Some practical examples are (1) the dehydrogenation of butylene to butadiene, which may further decompose to carbon and undesired products, and (2) the partial oxidation of naphthalene or ortho-xylene to phthalic anhydride, which may further oxidize to undesired products.

If reaction of A and B are each assumed to be first order and volume change is neglected, *at an effectiveness factor approaching unity* the ratio of rate of change of B to that of A, which may be termed the selectivity, is

$$-\frac{dc_B}{dc_A} = 1 - \frac{1}{S} \times \frac{c_B}{c_A}, \tag{5.2}$$

where S is the intrinsic selectivity *factor* k_1/k_2.

Integration of Equation 5.2 along the reactor gives

$$f_B = \frac{S}{S-1}(1 - f_A)[(1 - f_A)^{(1-S)/S} - 1], \tag{5.3}$$

where f_A is the fraction of initial A reacted and f_B is the fraction of initial A present as B (the remainder is C and unreacted A).

For the case of a porous catalyst operating *at a low effectiveness factor* ($\phi_L > 3$), Wheeler [382] shows that the ratio of the rate of change of B to that of A in a catalyst pellet at a point in a reactor is

$$-\frac{dc_B}{dc_A} = \frac{\sqrt{S}}{1+\sqrt{S}} - \frac{1}{\sqrt{S}}\frac{c_B}{c_A} \quad \text{(Type III selectivity, } \phi_L > 3). \tag{5.4}$$

Comparison of Equations 5.4 and 5.2 shows the effect of a low effectiveness factor upon the observed selectivity $-dc_B/dc_A$. Integration of Equation 5.4 along the reactor gives

$$f_B = \frac{S}{S-1}(1 - f_A)[(1 - f_A)^{(1-\sqrt{S})/\sqrt{S}} - 1] \quad \text{(Type III selectivity, } \phi_L > 3). \tag{5.5}$$

A plot of f_B versus f_A from Equation 5.3 compared to a plot of f_B versus f_A from Equation 5.5 shows the effect of a low effectiveness factor in altering the amounts of B formed. Figure 5.3 shows such a comparison for an intrinsic selectivity factor k_1/k_2 of 4.0. The maximum yield of B obtainable when a porous catalyst is operating at a low effectiveness factor is only about one-half that obtainable if the effectiveness factor approaches unity. Whenever the ratio of rate constants (intrinsic selectivity factor S) exceeds 1, the effectiveness factor with respect to A will always be less than that with respect to B, assuming the differences between the effective diffusivities to be negligible. Hence, at low effectiveness factors for both, the ratio of rate of reaction of B in a catalyst pellet to that of A will be greater than it would be on a plane surface, and thus a poorer yield of B is obtained. For intrinsic selectivity factors varying up to 10, the maximum yield loss due to a low effectiveness factor is reported to be about 50 per cent [382].

Although the effectiveness factor continues to drop as the Thiele modulus increases, the observed selectivity becomes independent of the modulus at values of ϕ_L exceeding about 3 ($\eta < 0.3$). The decrease in selectivity all occurs at effectiveness factors between unity and about 0.3. If a series reaction is being carried out on catalyst pellets of such size that the effectiveness factor is substantially less than unity but above

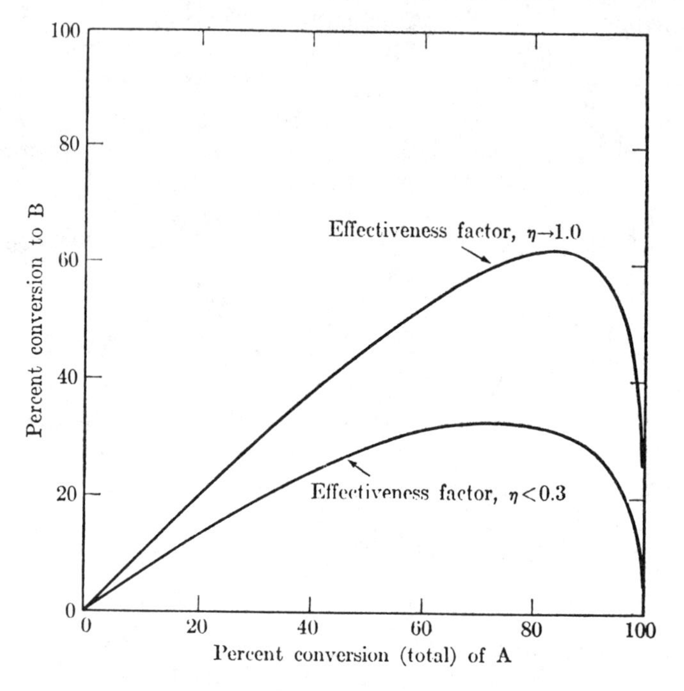

Figure 5.3. Effect of effectiveness factor on catalyst selectivity. Reaction of the type $A \xrightarrow{k_1} B \xrightarrow{k_2} C$. Calculated values for $k_1/k_2 = 4.0$. Wheeler [382, 383].

about 0.3, it should be possible to increase the yield of a desired intermediate substantially by subdividing the catalyst or altering the pore structure so as to increase the effective diffusivity. If, however, the effectiveness factor is well below 0.3, a large reduction in pellet size or large increase in effective diffusivity is required to achieve significant improvement in selectivity.

Other forms of selectivity may also be considered. For the case of two independent simultaneous reactions, termed Type I by Wheeler,

$$A \xrightarrow{k_1} \text{products and } B \xrightarrow{k_2} \text{products} \qquad \text{(Type I selectivity).} \quad (5.6)$$

The selectivity, for any value of η, is given by

$$-\frac{dc_B}{dc_A}=\frac{c_B\,k_2\,\eta_B}{c_A\,k_1\eta_A}. \tag{5.7}$$

At low effectiveness factors, η is inversely proportional to ϕ and Equation 5.7 becomes

$$-\frac{dc_B}{dc_A}=\frac{c_B}{c_A}\sqrt{\frac{k_2\,D_{B,\,eff}}{k_1 D_{A,\,eff}}} \quad \text{(Type I selectivity, low effectiveness factor).} \tag{5.8}$$

Equation (5.8) shows that it is possible for the selectivity $-dc_B/dc_A$ of two independent, simultaneous reactions in a porous catalyst to be either greater or less than it would be on a plane catalyst surface. If the ratio of effective diffusivities is less than the ratio of intrinsic rate constants (as in the case of the larger molecule being the more reactive), the selectivity ratio will drop as the effectiveness factors are decreased. However, if the more reactive molecules also have a much higher effective diffusivity, it would be possible for selectivity to be *increased* in a porous catalyst. This can easily happen with a molecular sieve type of catalyst, in which one reactant can be essentially excluded from the catalyst interior by its molecular size or shape (see Section 1.7.4).

If substantial pore-mouth poisoning occurs in a porous catalyst, with a low effectiveness factor the rate becomes limited by the rate of diffusion through inert pore mouths and the observed selectivity becomes

$$-\frac{dc_B}{dc_A}=\frac{c_B}{c_A}\sqrt{\frac{D_{B,\,eff}}{D_{A,\,eff}}} \quad \text{(pore-mouth poisoning).} \tag{5.9}$$

For parallel reactions of the type

$$A \begin{array}{l} \overset{k_1}{\nearrow} B \\ \underset{k_2}{\searrow} C \end{array} \quad \text{(Type II selectivity),} \tag{5.10}$$

the effectiveness factor does not affect the selectivity if both reactions are of the same order. However, Wheeler has pointed out [382, 383] that, if reaction 1 were first order and reaction 2 were second or higher order, then the observed selectivity in forming B will increase with decreased effectiveness factor. As the concentration of A drops toward the center of the catalyst pellet, the rate of reaction 2 will fall more rapidly than that of 1. If the desired product is B, and it is found that its yield is increased at lower pressures, then the above analysis suggests it should be improved further by using a catalyst operating at a low effectiveness factor, i.e., one with pellets of larger dimension or with smaller pores.

These theoretical analyses have been extended and generalized by Carberry [62] for Type III kinetics (Equation 5.1) for various combinations of mass transfer limitation in the bulk phase, pore diffusion, and diffusion in a macro-microporous structure. Ostergaard [237] has analyzed a series of consecutive first-order reactions, and gives the calculated conversion to the first intermediate product as a function of the Thiele modulus and several values of the selectivity factor. All of the above analyses assume that the exterior surface area of the catalyst particle is negligible compared to the total. At increasingly low effectiveness factors, however, the reaction becomes confined more and more to the outermost shell of the catalyst particle. Kramer [176] demonstrates that at very high values of the Thiele modulus the selectivity of a reaction on a porous catalyst should again approach that to be expected on a plane or nonporous surface, and he develops criteria to establish when this should occur. In practical situations, however, it would seem that this corresponds to such a high rate of reaction that the selectivity would be determined rather by the mass transfer characteristics to the catalyst surface from the bulk phase, which would presumably be the limiting step. A possible case might develop with reaction in ion-exchange resins in which the effective diffusivity was extremely low, or with single crystals of molecular sieves.

The selectivities produced by nonisothermal effects combined with complex kinetics can be discussed meaningfully only by individual groups of cases. Butt [52] has extended the nonisothermal treatment of Weisz and Hicks to Type III kinetics (Equation 5.1) and presents charts showing the effect of the Thiele modulus on the effectiveness factor and selectivity for various ratios of S and various values of β_1, β_2, γ_1, and γ_2, where the subscripts 1 and 2 refer to the two steps in the over-all reaction. The thermal characteristics of the second reaction, such as its enthalpy change and activation energy, can influence the temperature gradient through the catalyst pellet and thereby have a significant influence on the effectiveness factor with respect to the initial reaction of A. For isothermal conditions and a simple irreversible reaction, the effectiveness factor for reactant A is independent of the nature of subsequent reactions, except insofar as the products formed may affect the diffusivity of A or alter the intrinsic activity of the catalyst in some way. As is the case with other isothermal systems, the selectivity is constant throughout the entire highly diffusion limited region, which here corresponds to values of $\phi_{L,2}$ exceeding about 3. Hutchings and

Carberry [157] also report some computed results of temperature gradients on Type III kinetics with diffusion, and Ostergaard [239] discusses a special case of Type II kinetics.

Diffusion limitations are usually undesirable, under either isothermal or nonisothermal conditions, in the sense that the selectivity is usually less than would be achieved in the absence of diffusional limitations. A few exceptions to this generalization have been pointed out above for isothermal systems. Another exception occurs in Type III kinetics when the activation energy of the first reaction exceeds that of the second. Here it is possible for the yield of an intermediate to be enhanced by diffusional limitations coupled with a temperature gradient over that obtainable in the absence of diffusional limitations. Certain net endothermic systems may show improved selectivity under nonisothermal conditions over that for the isothermal case [52].

5.2.1 *Experimental Observations*

From a practical point of view, it will almost always be desirable to achieve effectiveness factors approaching unity, except for such specialized cases as the incorporation of molecular sieves into the catalyst specifically in order to achieve a desired exclusion of certain reactants from reaction sites. Uniform poisoning of a catalyst is a practicable method of minimizing diffusion limitations in some cases. A study by Weisz and Swegler [379] confirmed the conclusions of theory. They cracked cyclohexane to cyclohexene and thence to benzene on a chromia-alumina catalyst in a differential reactor, using three different catalyst sizes. As expected, the maximum yield of cyclohexene increased with decreasing particle size and if one applies a material balance to their results, one finds a greater degree of nonreported, products (and possibly coke) with the larger than with the smaller catalyst sizes. Other studies report more qualitative, yet significant, effects of diffusion limitations on selectivity. Johnson, Kreger, and Erickson [159] studied the catalytic cracking of gas oil on silica-alumina catalysts of several sizes, and reported that the selectivity to form gasoline increased with decreasing particle size. Gamid-Zade, Efimova, and Buzova [115] report studies of cracking on silica-alumina catalysts of various pore sizes of a petroleum distillate having a boiling point range of 350–500°C. The porosity of the catalyst was varied by changing the temperature of processing the catalyst. At 450°C, the highest selectivity to form gasoline was obtained when the catalyst pores were under 20 Å. For production of diesel fuel, optimum pore sizes were reported to be 20–50 Å.

An interesting application of the above principles to a practical catalyst problem is reported by Mars and Gorgels [199]. It is necessary to remove traces of acetylene from ethylene produced by thermal cracking before the ethylene can be used chemically for most applications. By adding sufficient hydrogen so the ratio of hydrogen to acetylene exceeds unity, the selectivity of hydrogenation of acetylene to ethylene over a palladium-on-alumina catalyst is essentially complete. Typical reaction conditions are 25 atm and 50–200°C. As long as acetylene is present, it is selectively adsorbed and hydrogenated. However, once it disappears, the undesired hydrogenation of ethylene takes place. The intrinsic kinetics are first order with respect to hydrogen and zero order with respect to ethylene or acetylene. Clearly, in the presence of diffusion limitations with respect to acetylene, ethylene hydrogenation will take place in the core of catalyst particles. Experimental studies by Mars and Gorgels showed that the problem could be solved in the various ways suggested by theory, e.g., the use of a large-pore catalyst, supporting the palladium in a thin outside layer on the catalyst, or by operating at sufficiently low temperatures that the rate dropped to a level corresponding to an effectiveness factor of essentially unity. Characteristically, these catalysts do not achieve maximum selectivity under plant conditions until they have been on stream for some days, and it appears that this is caused by a gradual poisoning from traces of such compounds as carbon monoxide. This may be an intrinsic kinetic effect, but the same behavior would be expected if reaction conditions were initially diffusion limiting and the poisoning reduced the reaction rate to bring the effectiveness factor up to near unity.

A number of curious and otherwise unexplainable catalytic phenomena under industrial processing conditions have been traced to effects produced by diffusion. In one highly endothermic reaction, the selectivity in a continuous packed-bed reactor gradually improved with time as the activity gradually decreased. The reaction was highly diffusion limited. It was discovered that a catalytic promotor, which was very slightly volatile, was being gradually displaced downward through the bed, but, more importantly, it was also being displaced toward the center of each catalyst pellet under the temperature gradient in each pellet set up by the endothermic reaction. The consequent lowering of activity in the outer portions of the catalyst raised the over-all effectiveness factor toward unity.

Catalyst activity can be profoundly affected by the local environment which can cause a change in the valence state, rate of poisoning, etc., so that under diffusion-limiting conditions the sites in the interior of a catalyst pellet may be subjected to more unfavorable conditions than those on the outside. An example is the commercial ammonia synthesis catalyst, which consists of iron oxide promoted with small amounts of substances such as Al_2O_3, K_2O, and CaO. One of the functions of the promoters is to maintain a large surface area during reduction with hydrogen to form the active catalyst and during the synthesis reaction itself. Nielsen and co-workers [227] show that water vapor formed during the reduction process accelerates the recrystallization of iron and lowers the ultimate surface area. A catalyst of higher surface area could be prepared by allowing the water vapor formed during reduction to escape more readily from the catalyst interior by reducing particle size or increasing the effective diffusivity, the latter by grinding and pelletizing the catalyst so as to produce a macro-micro pore-size distribution. The latter procedure significantly increases the effective diffusivity only if reduction is carried out at a pressure level at which diffusion is in the Knudsen or transition region where D_{eff} increases with increased pore size.

In continuous dehydrogenation reactions a substantial excess of hydrogen or steam is usually added to the reactant to delay or prevent the buildup of carbonaceous deposits on the catalyst. It has been observed in several cases that the rate of deposit formation in these reactions is decreased with larger-pore catalysts. Polymer precursors probably have increased difficulty in diffusing out of fine pores and in an extreme they may gradually block off access of hydrogen or steam, thus causing a self-accelerating process of formation of polymer and coke. This advantage of large-pore catalysts may, of course, be offset by their generally lower area and consequent lower activity per unit volume. An example of this effect is reported by Gardner and Hutchinson [116], who studied the hydrocracking of *m* terphenyl in a trickle-bed reactor on a variety of catalysts at 500 psig, 900°F and in the presence of a great excess of hydrogen. This is a model compound representative of an organic coolant used in nuclear reactors. The rate of coke formation on catalysts of 180 m^2/g surface area was four times greater than for catalysts 25 to 100 m^2/g. This could, of course, merely reflect increased rate of reaction due to greater area, but some of the catalysts eventually disintegrated, apparently because of the formation of polymer and swelling inside pores. Of a series of nickel catalysts, the only ones stable physically had surface areas below 100 m^2/g and nickel contents below 5 per cent. Even when physical disintegration of a catalyst does not occur under conditions in which deposits may gradually develop, it frequently appears that the fine pores of a catalyst become plugged fairly rapidly and only the larger pores contribute to the long-term catalyst activity. A catalyst with a macro-micropore distribution may show essentially the same long-term activity as a catalyst of much lower surface area containing only macropores whereas a high area catalyst containing only micropores may show rapid falloff of activity to an extremely low steady-state level.

The change in selectivity of a reaction upon altering the physical structure of a catalyst can also be used as a diagnostic test for the onset of diffusion limitations, as in a study by Newham and Burwell [225] of the hydrogenolysis of dicyclopropylmethane on a series of supported platinum and nickel catalysts and in certain studies with molecular sieve catalysts (Section 1.7.4). The principal problem here is to be sure that the change in selectivity is indeed caused by diffusion and not by a change in the intrinsic kinetics of the catalyst.

Wei and Prater [367] have made a fundamental attack on the problem of analyzing complex systems of first-order processes including the combination of a first-order reaction and diffusion, under both steady-state and transient conditions. The mathematics is quite complex and involves presentation of kinetic information in geometrical form which is then treated by use of linear algebra and matrix operations. The approach is presented in detail in a lengthy introductory paper [367], and some of its applications are pursued in a series of subsequent publications [366]. Many of the applications and results of calculations can be appreciated, however, without the necessity to master the mathematics involved. For example, three-component reacting systems, both reversible and irreversible, are treated in detail to show the effects of catalyst diffusivity on the reaction paths followed.

5.3 Regeneration of Coked Catalysts

In many hydrocarbon reactions such as catalytic cracking, reforming, desulfurization, and various dehydrogenations, carbonaceous deposits of so-called "coke" are gradually formed on the surface of the catalyst. The continuing accumulation of these deposits reduces the activity of the catalyst to the point that it must be regenerated. This buildup of coke may become significant in a few minutes, as in catalytic cracking or dehydrogenation of butane to butadiene, or only over a period of months, as in catalytic reforming. Regeneration of the catalyst is accomplished by burning off the coke with air, a mixture of air and nitrogen, or a mixture of air and steam. If regeneration is required often, a continuous process must provide a facility separate from the reactor for continuous burnoff, which is generally of the same type as the primary reactor, e.g., a fluidized bed or moving bed. Alternatively, a cyclic fixed-bed process may be designed in which reaction and coke burnoff steps follow in sequence, separated by purge steps to avoid the formation of explosive gas mixtures. An example is the Houdry butadiene-from-butane process. If regeneration is only required at intervals of several months or longer, a fixed-bed unit is generally shut down for the period of time required for regeneration and then put back on stream. If this interruption is too lengthy, or regeneration is difficult to achieve in the reactor vessel, the coked catalyst may be quickly replaced with fresh catalyst and the regeneration carried out elsewhere. The method of regeneration used depends upon many factors; these include the relative availability of steam or inert gas, the sensitivity of the catalyst to elevated temperature, to steam or carbon dioxide, the design of the reactor and ability to remove heat during the regeneration. The use of air or air plus an inert gas is by far the most common regeneration method and almost all of the fundamental information on regeneration is concerned with the reaction of coke and oxygen. In fixed-bed reactors, the burning takes place largely within a burning zone which moves slowly through the reactor, the temperature along the bed varying greatly with position and with time. The following discussion will focus primarily on the processes occurring in a single pellet or in a differential reactor.

If the effectiveness factor for the primary catalytic reaction is essentially unity and the catalyst is uniform in intrinsic activity, the coke will be initially distributed quite uniformly through the catalyst. At sufficiently low temperatures of regeneration, the oxygen gradient through the pellet will be negligible, and the effectiveness factor for the regeneration reaction approaches unity. The coke concentration remains uniform throughout the pellet until completion of the oxidation reaction.

At sufficiently high temperatures, oxygen reacts as fast as it is transported to the carbon. In a sphere, reaction occurs solely at a spherical interface which moves progressively to the center, and the rate is limited by the rate of diffusion of oxygen through a shell of carbon-free porous solid. These facts may be visually demonstrated by cutting a pellet in half after partial removal of coke, or by submerging it in a liquid of the same refractive index, e.g., silica-alumina beads in benzene. Under highly diffusion-controlled conditions, one sees [369, 374] a black sphere, surrounded concentrically by a white or light-gray spherical shell from which all the coke has been removed. Under intermediate conditions, a progressive darkening toward the center is visible.

A substantial number of reports are available on the rate of combustion in air of carbonaceous deposits in porous catalysts, both of the intrinsic kinetics and under diffusion-controlled conditions; however, in many of the studies, particularly earlier ones, operating conditions were poorly controlled, the reactor was not truly differential, or measurements were made in the transition zone between intrinsic kinetics and diffusion control. All these factors make interpretation difficult. One of the most clearly defined studies is by Weisz and Goodwin [370, 374, 375].

5.3.1 *Intrinsic Kinetics: Nature of Coke Deposits*

An atom of carbon in graphite occupies about 4 $Å^2$ of cross-sectional area; if each atom of carbon in "coke" occupies the same area, a monatomic layer of carbon on an oxide-type catalyst or support would comprise about 5 wt % carbon for each 100 m^2/g of surface area [375]. Studies by Weisz and Goodwin [375] on silica-alumina catalysts of about 250 m^2/g surface area indicated that up to about 6 wt % carbon, the rate of combustion was proportional to the amount of carbon present; i.e., all atoms of the coke were presumably equally accessible to oxygen. At higher coke contents (e.g., 7–20 per cent C on silica-alumina) the rate for the first 50 per cent or so of reaction was less than proportional to the amount of carbon present, indicating that some of the carbon atoms were initially inaccessible, but in the latter part of the reaction the rate again became first order with respect to total carbon present. The same type of behavior at high coke contents has been reported by others. The character of these deposits varies substantially with the conditions under which they were formed. In some cases, they are essentially an ill-defined high-molecular-weight polymer, most of which can be removed by washing with a solvent. Appleby, Gibson, and Good [8] made a detailed characterization of the coke formed in catalytic cracking and show that it consists of large aggregates of polynuclear aromatic molecules comprising essentially condensed systems of fused aromatic rings, plus strongly adsorbed portions of products of the reaction, plus coke deposits remaining unregenerated from a previous cycle. Other studies, as by electron microscopy and differential thermal analysis, have also been reported [131, 132]. The carbon or coke structure is evidently of considerable complexity and varies substantially with conditions of formation and subsequent treatment, but X-ray diffraction shows that a substantial portion is in a pseudographite form. Most investigators of the regeneration reactions have pretreated their coked catalyst at elevated temperatures in an inert atmosphere before combustion in order to eliminate most of the volatile matter present or to further decompose the coke and thereby obtain more reproducible results (see later). In most laboratory studies the coke has been deposited from a hydrocarbon vapor under non-diffusion-limiting conditions, which at moderate carbon loadings seems to provide a true monolayer. However, in a recent study by Massoth [204] a 75 per cent silica–25 per cent alumina catalyst of 349 m^2/g was coked by first soaking it in a furnace oil and then subjecting it to 425 °C under a hydrogen pressure of 750 psig. After further pretreatment, the deposit had an average composition of $CH_{0.4}$ and the carbon content on the catalyst was 9.6 wt %. The oxidation characteristics of this material indicated that, although the carbon was well dispersed through the catalyst, it seemed to be present initially in the form of particles about 100 Å in diameter rather than as a monolayer. The coke as deposited by any of these methods generally contains considerable hydrogen, the empirical formula usually being between CH_1 and $CH_{0.5}$. The hydrogen is removed preferentially in the early stages of the reaction; the first material reacted has an empirical formula of about CH_2 [204]. Thus, the latter part of the reaction is essentially that of a carbon skeleton on the catalyst.

Carbon Gasification Kinetics. Figure 5.4 is an Arrhenius plot of the first-order reaction-rate constants for combustion of monolayer coke deposits on various kinds of supports. The rates reported are usually (although not always) based on the amount of carbon gasified. The combustion of carbon as such in its various forms has been extensively studied [88, 99, 285, 358] and will not be reviewed here. The observed intrinsic activation energies all fall between 35 and 40 kcal/g-mol, but adsorption of oxygen and desorption of products affect the kinetics in ways which vary substantially with temperature, pressure, and nature of the carbon. The intrinsic rate of reaction is strongly affected by the form of the carbon and the nature of the impurities present. In view of this, the degree of agreement among burning rates reported by several different investigators for monolayer carbon deposits on various catalysts, within a factor of about 2, is remarkably good. This holds for situations in which the catalyst or support does not contain transition metals or other substances which catalyze the oxidation of carbon. The line on the figure for this situation represents data reported by Weisz and Goodwin [375] for coke burnoff from silica-alumina, silica-alumina "Durabead," silica-magnesia, Fuller's earth (an adsorbent clay), or Filtrol 110 (a cracking catalyst derived from natural clay). There was no effect of support on the rate. Johnson and Mayland [160] also reported

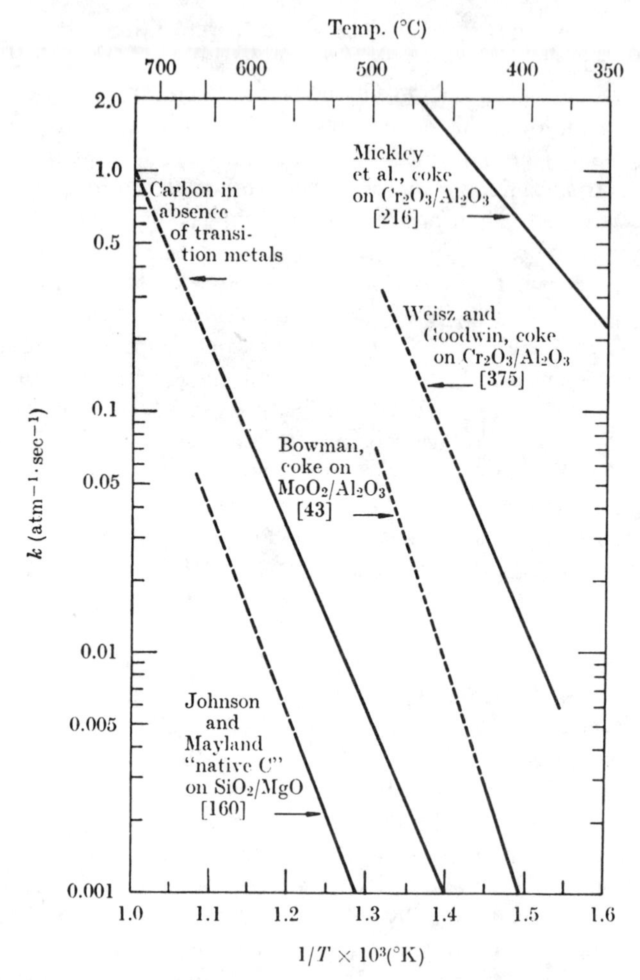

Figure 5.4. First-order reaction-rate constants for oxidation of coke deposits on various catalysts and supports. Dashed lines indicate extrapolated data.

in an earlier study that the specific burning rates of carbon on silica-alumina, silica-magnesia, clay, silica gel, and cracking catalysts were approximately the same, although detailed information about the kinetics on the various supports was not developed. The carbon burnoff rate is independent of the source of the coke provided that it is freshly deposited from the vapor phase in not-excessive amounts, and is "seasoned" (see below). The same results were obtained [375] on silica-alumina with deposits from laboratory cracking of gas oil, of cumene, naphtha, or propylene. The equation for the first-order rate constant is given by [375]

$$k = 1.9 \times 10^8 \exp(-37\,600/\boldsymbol{R}T)\text{sec}^{-1} \cdot \text{atm}^{-1}, \tag{5.11}$$

where the expression for the rate of reaction is

$$-(1/V_c)(dn/dt) = k \cdot P_{O_2} \cdot c_C$$

and c_C is the moles of carbon present per cm^3 of pellet.

Several other investigators show results that agree closely with Equation 5.11. Hagerbaumer and Lee [130] in 1947 reported on the combustion of coke from a synthetic silica-alumina bead catalyst about $\frac{1}{8}$ in. in diameter and having an average pore diameter of 47 Å. After removal of the first 40 per cent of the carbon, the reaction was first order with respect to the remaining carbon content. The activation energy was 35 kcal /g-mol and above about 480°C it fell off, indicating the onset of diffusion control. Hagerbaumer and Lee's data essentially agree with those of Weisz and Goodwin even though a considerable amount of empirical correlations seemed to have been required in analyzing the data. Pansing [245] likewise reported data on regeneration of a cracking catalyst and an activation energy of 35 kcal/g-mol which agree with the above two studies. Massoth's [204] study (425–480 °C and O_2 concentration of 0.04 to 0.21 atm) also showed an activation energy of about 40 kcal/g-mol and rate data, interpreted as burning of small clumps of carbon, which also essentially agree with the above.

Adel'son and Zaitova [1] reported studies of coke burning from 0.33-cm-diameter silica-alumina cracking catalyst spheres at 450–500°C and 0.21 atm oxygen. They also reported the reaction to be first order with respect to carbon with an activation energy of 35 kcal/g-mol, but their rates were about two-thirds of those reported by Weisz and Goodwin. Equation 5.11 may thus be taken as representative of the intrinsic combustion rate of monolayer carbon on a noncatalytic surface. There is some controversy concerning the order of the reaction with respect to oxygen under various conditions, but most investigators indicate that it is nearly proportional to the first power [358], in the range from about 0.1 to 1 atm of oxygen and at the temperatures of interest in catalyst regeneration.

These rates are surprisingly close to those reported for oxidation of graphite. Effron and Hoelscher [98] studied oxidation rates of four types of graphites at temperatures of 420–533°C, mostly in 1 atm of pure oxygen. The rates of reaction per unit area varied by a factor of about 5 between the most active and the least active graphites, but the activation energy was 35.5 kcal/g-mol in each case. The rate on the most active graphite agrees closely with Equation 5.11. An earlier study of graphite oxidation by Gulbransen and Andrew at 0.1 atm oxygen is reported by Weisz and Goodwin [375] to agree closely with their data at 0.21 atm and represented by Equation 5.11. The results of Essenhigh, Froberg, and Howard [99] on a pure carbon are, however, higher by a factor of about 10.

The analysis of other studies as summarized in Figure 5.4, in light of the above, is instructive. The exceptions to the above generalization seem to fall into three categories. The presence in the porous structure of transition metals or other catalysts for carbon combustion greatly accelerates the rate of carbon oxidation. A chromia-alumina catalyst shows rates two to three orders of magnitude greater than that observed on silica-alumina, although the activation energy remains the same [375]. Even a trace of chromia (e.g., 0.1 per cent) in silica-alumina increases the rate in fresh catalyst by a factor of about 4, although the rate drops to that of coke on silica-alumina alone after about three days in a commercial cracking installation or after steam treatment. Apparently this is because the chromium is converted from the Cr^{6+} form to Cr^{3+}. Bowman's studies were on a molybdena-alumina catalyst that evidently had catalytic activity for carbon oxidation. Platinum-on-alumina reforming catalysts also are catalytic for carbon oxidation and can be rapidly regenerated at relatively low temperatures and low oxygen concentration. The study of Mickley, Nestor, and Gould [216] was on a Houdry chromia-alumina dehydrogenation catalyst. The rates are one to two orders of magnitude higher than those reported by Weisz and Goodwin for a similar chromia-alumina catalyst and probably represent primarily the removal and oxidation of unstripped volatile hydrocarbon remaining in the catalyst pores rather than carbon combustion as such. The regeneration was separated from the coking by only a 30-sec nitrogen purge, representative of commercial practice in the fixed-bed Houdry cyclic process for production of butadiene from butane. The method of purging and treatment before combustion can greatly affect the initial burning rates, and this purge time is much shorter than that used by most other investigators (see below). At higher temperatures (570–640°C), where the reaction was at least partly diffusion controlled, the observed rate of reaction actually dropped by a factor of about 4 from the maximum observed (at about 475°C). This suggests that at the intermediate temperature range a considerable amount of volatile material was being burned outside the catalyst pores.

Klimenok, Andreev, and Gordeeva [171] were able to remove some 60 per cent of their "coke" deposited by cracking hexane on an Si-Al catalyst by evacuation at 350–400°C; the initially rapid rates reported by various investigators may represent desorption of volatile hydrocarbons, further cracking combined with desorption, or the inititial rapid reaction of hydrogen present in the coke. Steam formed in the catalyst pores in the initial part of the reaction of hydrogen in the "coke" may also provide a mechanism to accelerate the gasification of the carbon. To eliminate these effects, Weisz and Goodwin heated their samples in an inert atmosphere for at least 15 min at 340°C or the temperature of the burning reaction, whichever was higher. Haldeman and Botty [131] likewise aged their coke, at 565°C, for periods of up to 1 hr and other investigators have used similar procedures. Johnson and Mayland in studies of purging time at 540°C on silica-magnesia, after carbonization with benzene, showed a burning rate after about 2 min of purging approximately 15 times greater than that after 1 hr of purging.

Johnson and Mayland reported data for burnoff of "native carbon" at the 0.3 wt % level from a silica-magnesia catalyst from a pilot plant regenerator; these are shown in Figure 5.4. For other commercially regenerated catalysts such as various clays and silica-alumina at the 0.3 per cent carbon level, they likewise reported burning rates for native carbon roughly one-half of those obtained when the carbon was freshly deposited and aged. In some commercial operations, all catalyst particles are probably not completely regenerated at each cycle, and there is an opportunity for multilayered carbon deposits to build up even at low over-all coke levels. However, Weisz and Goodwin report no difference between carbon burnoff rates for carbon deposited in the laboratory and that produced in commercial refinery cracking on a moving-bed reactor. Freshly deposited carbon on aged catalysts behaves like freshly deposited carbon on virgin catalysts; i.e., alterations in pore size and physical structure of the catalyst upon aging apparently do not affect the intrinsic kinetics as such.

In order to compare the various studies in Figure 5.4, some of the data have been extrapolated using activation energies reported by the investigators. The Johnson and Mayland studies were at temperatures of 510–565°C, the Weisz and Goodwin studies of noncatalytic burning covered temperatures of 450–600°C, on chromia-alumina about 370–440°C. Mickley's study reflected intrinsic kinetics up to about 470°C and Bowman's study began to become diffusion controlled above about 425°C. Hagerbaumer and Lee's data were for temperatures up to about 480°C, above which diffusion became significant.

Johnson and Mayland also report some limited information on the effect of steam addition to air on the rate of carbon gasification.

5.3.2 *Diffusion-Limited Reaction in the Gasification of Carbon*

At sufficiently high temperatures, the rate of reaction becomes limited by the rate of diffusion of oxygen through the carbon-free pores in the burned-off annulus, or shell, to the interface where reaction is occurring. The situation is analogous to reaction occurring in a porous catalyst with pore-mouth poisoning in which the poisoned area increases in proportion to the amount of reaction occurring. The burnoff rate may then be expressed by [377]

$$\tfrac{1}{2}(1 - y^{2/3}) - \tfrac{1}{3}(1 - y) = \frac{v D_{\text{eff}} c_s t}{R^2 c_C}, \tag{5.12}$$

where y is the fraction of initial carbon remaining at time t, c_s is the bulk concentration of oxygen in the air or combustion gas, c_C is the initial molar concentration of carbon on the catalyst pellet in mol/cm^3 pellet. v is the stoichiometric coefficient, number of C atoms removed per oxygen molecule diffusing into the catalyst pellet; this is determined by the C/H ratio of the coke and the CO/CO_2 ratio *leaving the pellet.* Equation 5.12 assumes that the deposit of carbon is initially uniform throughout the pellet, as would usually be the case if the primary reaction in which the carbon was formed involved no significant diffusion limitations.

The equivalent expression for flat plate geometry (sealed on one side) is

$$\frac{(1 - y)^2}{18} = \frac{v D_{\text{eff}} c_s t}{R^2 c_C}, \tag{5.13}$$

where the plate thickness $L = \frac{1}{3}R$. The two expressions are nearly identical for values of y down to about 0.6 [63]. Carberry and Gorring [63] suggest as a working criterion for shell-type reaction the condition in which the reactant concentration drops to $e^{-4} = 0.0183$ of the external concentration within a distance of 2 per cent of the flat plate thickness. For first-order reaction the thickness of the reaction zone thus defined is independent of its position within the catalyst particle, and this criterion corresponds to $\phi_L > 200$, where ϕ_L is expressed in the usual manner. They also present curves showing how the fraction of carbon reacted varies with time for various degrees of bulk mass transfer limitation, and for three degrees of approach to the shell-type model, the latter represented by a Damköhler number, $N_{\text{Da}} = k_s R / D_{\text{eff}}$. Here

k_s is the intrinsic first-order reaction-rate constant (cm/sec). $N_{Da} \rightarrow \infty$ corresponds to shell-type reaction. The same mathematical expressions also apply to the deposit of a poison on a catalyst by a first-order process. The various analyses assume pseudo steady state throughout the reaction, an assumption that seems to be generally valid [193] except for the very beginning (see below) and very end of reaction.

For coke burnoff from homogeneous silica-alumina bead catalysts of typical radius (0.2 cm), diffusion of oxygen begins to become rate limiting at about 480°C. The above shell model, with sharp interface between the burned region and unburned coke, represents the situation at temperatures above about 540°C [370]. Equation 5.12 suggests several significant conclusions concerning the regeneration of coked catalyst pellets at high temperatures. For a fixed oxygen concentration outside the pellet, the time required for a given degree of regeneration, e.g., 85 per cent burnoff of the carbon, is directly proportional to the amount of carbon initially present in the pellet. The time required for a given per cent removal of carbon increases with the square of the pellet radius and is inversely proportional to the effective diffusivity. The latter is particularly noteworthy since the effective diffusivity of homogeneous silica-alumina bead catalysts can be increased by a factor of as much as 10 by incorporating fine powder in the forming solution before gelation [374], thus producing a wide pore-size distribution. The rate of burning is directly proportional to the oxygen concentration in the regenerating gases but since the reaction is diffusion controlled, the rate will not increase significantly as temperature is increased. The intrinsic activation energy for carbon burnoff is about 37 kcal/g-mol, but when diffusion becomes appreciable the apparent activation energy will drop to a value of the order of 1–2 kcal/g-mol, corresponding to diffusion through inert pores. Note that this drop in activation energy is analogous to that observed with progressive pore-mouth poisoning of catalysts, but is different from a uniform catalyst operating at a low effectiveness factor. In the latter, as temperature is raised, and effectiveness factor drops, the apparently activation energy also drops, but to a limiting value of about one-half that of the intrinsic activation energy.

If the individual pellets in a catalyst charge vary substantially in effective diffusivity from one to another, as might be caused by varying porosity or pore-size distribution, some pellets may be incompletely regenerated and thus, during repeated cycling, would build up a coked core blocking off the activity of much of the catalyst area. The left-hand side of Equation 5.12 shows that 50 per cent of the coke will be removed in about the first 12 per cent of the total time required for complete regeneration, and that slightly over half of the total time is required to remove the last 15 per cent of the carbon.

The validity of Equation 5.12 has been well established by an excellent series of burnoff studies at 700°C under differential reactor conditions by Weisz and Goodwin [374]. The theoretical function relating burnoff to time was followed closely by the actual results for the time intervals reported, which corresponded to the removal of the last 50 per cent down to 2 per cent of the initial carbon present (values of y from 0.5 to 0.02). The time for 85 per cent burnoff, taken as a convenient measure, was proportional to initial carbon levels over the range of 1 to 4 wt % and varied as the square of the diameter for beads from 0.3 to 0.5 cm diameter but fixed initial coke content, both of which are in accord with theory.

Coke burnoff data analyzed by Equation 5.12 provide a method of measuring effective diffusivities in a radial direction in a catalyst pellet. A particularly interesting series of studies [374] showed that, for a wide variety of bead catalysts, the ratio of the effective diffusivity from a burnoff measurement to that measured by hydrogen-nitrogen diffusion at 20°C *through* the pellet was constant. A twenty-five-fold variation in diffusivity was represented by the various beads, which included homogeneous silica-alumina, commercial silica-alumina containing 0.15 wt % chromia, both fresh and after some thermal catalyst damage, as well as chromia-alumina and silica-alumina with a wide pore-size distribution. The principal uncertainty in calculating a value of D_{eff} from Equation 5.12 concerns the proper value of v, which depends on the C/H ratio of the coke and the CO_2/CO ratio in the gas leaving the pellet. Most of the hydrogen reacts during the initial part of the combustion, and this period is not amenable to data reduction in any event, so this factor is unimportant. The CO_2/CO ratio however, varies with reaction temperature, depth of penetration of the reaction zone, and catalytic effects on the further reaction of CO with oxygen diffusing into the pellet (see below). If v is 2 (all carbon leaves pellet as CO), values of D_{eff} agree closely with those measured by hydrogen diffusion and corrected to 700°C and diffusion of oxygen assuming Knudsen diffusion (Equation 1.33). If all the carbon is assumed to leave the pellet as carbon dioxide, which seems to be somewhat closer to what actually happens, D_{eff} from Equation 5.12 is about double that predicted from flowthrough measurements with hydrogen. The agreement is nonetheless remarkable, considering the range of catalysts investigated. It shows that at least this group of catalysts must be essentially isotropic in structure, and validates the extension of diffusion measurements *through* this type of porous structure to reaction conditions in which diffusion is in-and-out in a radial direction.

The ratio of CO_2/CO formed at the actual coke site seemed to be the same as that found in the burning of carbon, and it decreased with increased temperature. As the CO diffuses outward, however, it reacts with oxygen diffusing in, in a region devoid of carbon, so the ratio of CO_2/CO in the gas leaving the pellet increases as the reaction zone moves progressively toward the pellet center and, averaged through the regeneration, increases with pellet size. The incorporation of a trace of chromia in the silica-alumina catalyzes this reaction, long after the chromia has lost its activity to catalyze the primary reaction of carbon with oxygen. Weisz makes an analysis of these effects [370] which shows good agreement with experiment. Toward the end of the regeneration, as much as 80 per cent of the CO originally formed at the coke site can be converted to CO_2 before escaping from silica-alumina pellets of typical size, even when the pellet does not contain chromia to accelerate the reaction to CO_2.

Various other studies of regeneration of coked catalysts in air indicate conditions under which diffusion of oxygen begins to become rate limiting. A pelleted clay catalyst with large pores (e.g., half exceeding 2000 Å in diameter) has a higher diffusivity than homogeneous silica-alumina beads; oxygen diffusion was found not to be rate limiting [88] even at 650°C. With the fine particle sizes used in fluidized-bed operations, oxygen diffusion is not rate controlling on silica-alumina or a similar catalyst, at least up to 700°C [369, 372]. Regeneration of a molybdena-alumina cracking catalyst of particle dimensions of the order 1 cm showed both diffusion and chemical reaction to be important in the range of 430–540°C and to be diffusion controlled above 540°C [43]. Regeneration of a molybdena-alumina hydroforming catalyst consisting of $\frac{3}{16}$-in. cylindrical pellets of 90 m^2/g surface area, at about 505°C with air at 1.7 atm, was apparently diffusion controlled [285]. Regeneration of a 20 per cent chromia-alumina catalyst $\frac{1}{8}$ in. diameter by 0.17 in. long and about 13 m^2/g began to become diffusion controlled above about 515°C [216]. Adel'son and Zaitova [1] report analyses of internal and external diffusion effects in carbon burnoff from catalysts at various air flow rates, oxygen concentrations, and at temperatures of 450–750°C. The conditions under which diffusion control will begin to become significant in coke burnoff (e.g., $\eta \lesssim 0.95$) can be estimated by the same methods used for catalytic reactions. Example 4.7 gives an illustrative calculation, which shows close agreement with experiment. The effects of diffusion limitations on *formation* of carbonaceous deposits are discussed in Section 5.1.

The fact that the hydrogen present in the coke (together with some of the carbon) is removed rapidly by oxidation, leaving behind a skeletal carbon, means that there is a range of operating conditions in which the hydrogen-removal reaction is limited by diffusion of oxygen into the catalyst, but the removal of the residual carbon is controlled by the intrinsic kinetics. This seemed to be the case in the study by Massoth [204] at 425–480°C and oxygen concentrations of 0.04 to 0.21 atm, on coked 10-20 mesh (about 2 mm diameter) silica–25 per cent alumina. Some sets of circumstances can, at least in theory, lead to a diffusion process with a double moving boundary, an inner one between coke and carbon, and an outer one between carbon and regenerated surface.

A hot spot travels through a fixed bed during regeneration, and it is vital that this moving maximum temperature not exceed that at which the catalyst loses activity. In the regeneration of heat-sensitive coked catalysts in a fixed bed, it is common practice to start with a gas stream containing a low concentration of oxygen and then later to switch to air at higher temperatures. It is sometimes observed that a hot spot travels through the bed during each of the two sets of regeneration conditions. This is consistent with the above concept of the regeneration process as comprising two relatively distinct groups of reactions differing substantially in their intrinsic reactivity.

Temperature gradients within the catalyst particle during regeneration are negligible during the quasi-steady state portion of the regeneration even with diffusion-controlled conditions, but Bondi, Miller and, Schlaffer [41] show that during the first 10 msec or so of regeneration of a fluid cracking catalyst, temperature differences of up to 200°C or more can develop momentarily between an individual catalyst particle and its surroundings if it contains appreciable amounts of highly reactive material such as unstripped gas oil and if the reaction is controlled by diffusion of oxygen to and within the particle. These two requisites may readily occur when a fresh catalyst is introduced continuously into a fluid cracking reactor-regenerator system operating with so-called "equilibrium catalyst," whose activity and surface area is generally much less than that of the fresh catalyst. A similar type of analysis by Zhorov *et al.* [398] leads to a similar conclusion — that the temperature of a catalyst particle may increase by 150–200°C for a few seconds at the start of regeneration. Luss and Amundson [195] present a theoretical study of the stability of a batch fluidized bed and Winegardner and Schmitz [392] of the stability of a reaction on a single sphere in a stagnant medium. Many of the analyses of catalyst poisoning (Section 5.1) apply also to coke gasification.

List of Symbols

Some symbols used only once are defined at point of use and are not included in this list.

- a ratio of outside area of catalyst pellets in packed bed to reactor volume (cm^2/cm^3)
- A_p outside surface area of catalyst particles per unit volume of slurry (cm^2/cm^3)
- A'_p outside surface area of single catalyst particle (cm^2)
- A_B total interfacial area of gas bubbles in unit volume of slurry (cm^2/cm^3 of expanded slurry)
- B modulus used for reversible reactions, Equation 4.12. Reduces to $Kp_{A,s}$ for irreversible, first-order reaction
- c concentration (g-mol/cm^3); c_1 for species 1; c_0 for concentration in the bulk gas phase; c_s for concentration at outside particle surface; c_t for concentration or molar density of mixture; c_e for equilibrium concentration; c_L for concentration in bulk liquid; c^* for saturation concentration of gas in liquid (Chapter 2); c_l for center of particle; c_s for local surface concentration inside particle (Section 1.7.3); c_g for local gas phase concentration (Section 1.7.3). $-dc_B/dc_A$ is the selectivity (Section 5.2).
- c_p heat capacity (g-cal/g · °C)
- C modulus used for reversible reactions, ratio of equilibrium partial pressure to that at outside surface of catalyst (Section 4.4)
- d_I impeller diameter (cm)
- d diameter of pore (cm)
- d_p diameter of particle (cm); d_B for diameter of gas bubble
- D coefficient of molecular diffusion (cm^2/sec)

D_{12} diffusion coefficient for bulk diffusion of species 1 in a mixture of species 1 and 2 (cm²/sec); D_{1m} for diffusion of species 1 in a multicomponent mixture (see Equation 1.15)

D_K Knudsen diffusion coefficient for a straight round pore (cm²/sec)

D_{eff} effective diffusion coefficient for a porous solid, based on *total* cross section normal to direction of diffusion; equal to ratio of flux to concentration gradient,

$$\frac{\text{g-mol/sec} \cdot \text{cm}^2 \text{ of solid}}{\left(\dfrac{\text{g-mol}}{\text{cm}^3 \text{ fluid}}\right)\left(\dfrac{1}{\text{cm of solid}}\right)} = \frac{\text{cm}^3 \text{ fluid}}{\text{cm solid} \cdot \text{sec}} = \text{cm}^2/\text{sec}$$

$D_{12,\text{eff}}$ effective diffusion coefficient for a porous solid under conditions of bulk or ordinary diffusion, based on *total* cross section of porous solid normal to direction of diffusion (cm²/sec); $D_{12,\text{eff}} = D_{12}\theta/\tau$

$D_{K1,\text{eff}}$ effective Knudsen diffusion coefficient for species 1 in a porous solid (cm²/sec); $D_{K1,\text{eff}} = D_{K1}\theta/\tau$

D_s surface diffusion coefficient (cm²/sec) (Equation 1.44)

D_T dispersion coefficient, in packed beds (cm²/sec) (Chapter 2)

E activation energy for reaction (g-cal/g-mol); E_D, activation energy for diffusion

$\mathbf{E}$ modulus used for complex reactions (Equation 4.8)

f measure of importance of mass transfer to pellet surface $= -(c_0 - c_s)/c_0$

f area availability factor (Figure 2.1)

g acceleration due to gravity (cm/sec²)

G mass velocity of fluid (g/sec · cm² of total or superficial bed cross section normal to mean flow); G_M for molal velocity, $= G/M$

h heat transfer coefficient (g-cal/sec · cm² · °C). If h is given in btu/hr · ft² · °F, multiply by 1.36×10^{-4} to convert to preceding units.

ΔH enthalpy change on reaction (g-cal/g-mol)

H gas holdup (cm³ of gas/cm³ of expanded liquid or slurry)

j_D $\dfrac{k_c\rho}{G} N_{Sc}^{2/3} = \dfrac{k_g P}{G_M} N_{Sc}^{2/3}$

j_H $\dfrac{h}{c_p G} N_{Pr}^{2/3}$

J diffusion flux relative to plane of no net molar transport (g-mol/sec · cm²)

$\boldsymbol{k}$ Boltzmann constant

k thermal conductivity of fluid (g-cal/sec · cm · °C)

k_c mass transfer coefficient (cm/sec); k_c^* for free-settling particle; k_L for gas bubbles to liquid; k_{Ls} for transfer through liquid film in trickle-bed reactor.

k_G mass transfer coefficient (g-mol/sec · cm² · atm); $k_G = k_c/RT$

k_s intrinsic first-order reaction-rate constant for surface reaction (cm/sec); k_s' for reaction of order m ((cm^3 of fluid)^m/(cm^2 of catalyst surface)(g-mol)$^{m-1}$ · sec)

k_v intrinsic first-order reaction-rate constant per unit volume of catalyst pellet (cm^3 of fluid/cm^3 of catalyst · sec = sec^{-1}); k_v' for reaction of order m ((cm^3 of fluid)^m/(g-mol)$^{m-1}$ · sec · cm^3 of catalyst volume); $k_v' = k_s' S_v = k' \rho_p S_g$

K parameter used for complex or reversible reactions (Chapter 4, Equations 4.6 and 4.13)

K_i adsorption equilibrium constant for ith species in Langmuir-Hinshelwood expression for surface reaction rate (atm^{-1})

K_e equilibrium constant for reaction

L ratio of catalyst volume to outside surface area through which reactant has access (cm). As used in ϕ_L (Equation 3.22) and Φ_L (Equation 3.32), L is specifically the thickness of an infinite plate sealed on one side and on the ends.

m order of reaction

m catalyst loading in slurry reactor (g catalyst/cm^3 of slurry as expanded by gas flow) (Chapter 2)

M molecular weight

$-dn/dt$ rate of reaction (g-mol/sec); $-(1/V_c)(dn/dt)$ is the *observed* rate of reaction (g-mol/sec · cm^3 of catalyst pellet).

N diffusion flux relative to stationary coordinate system (g-mol/sec · cm^2)

N_{Pe} Peclet number, $=d_p U/D$; N_{Pe}^* for free-settling particle (Equation 2.45); in packed beds, $N_{Pe} = d_p u/D_T$

N_{Pr} Prandtl number, $= c_p \mu/k$

N_{Re} Reynolds number, $= d_p G/\mu$

N_s stirrer speed (rps)

N_{Sc} Schmidt number, $= \mu/\rho D$

N_{Sh} Sherwood number, $k_c d/D_{1m}$. Sometimes termed the Nusselt number for mass transfer.

p partial pressure (atm)

P total pressure (atm)

Q rate of formation of species (as by chemical reaction) (g-mol/sec · cm^3)

r reaction rate (g-mol reactant/sec · g catalyst)

r distance from center of sphere (cm) (Chapter 3)

r_e equivalent radius of pore (cm), $=2V_g/S_g$ (Equation 1.32)

R radius of sphere (cm)

$\boldsymbol{R}$ gas constant, $=1.987$ g-cal/g-mol · °K or 82.057 cm^3 · atm/g-mol · °K (Set as R in Chapter 1)

S intrinsic selectivity factor, $=k_1/k_2$ for first-order reactions; S_{obs} for observed selectivity $= -dc_B/dc_A$

S_g total surface area of porous solid (cm²/g) (as measured, for example, by B.E.T. method)

S_v total area of catalyst per unit volume of catalyst pellet (cm²/cm³) $= \rho_p S_g$

t time (sec)

T temperature (degrees Kelvin or Centigrade); T_s at outside surface of particle; T_0 for bulk gas phase; T_c for critical temperature

u mean molecular velocity (cm/sec); $u_1 = N_1/c_1$; $u_m = N/c_T$ (Chapter 1)

u average linear velocity in particle interstices in a packed bed (cm/sec) (Chapter 2). $u = U_0/\varepsilon$

u_m velocity of plane of no net molal transfer (cm/sec)

u_w velocity of plane of no net mass transfer (cm/sec)

U fluid velocity (cm/sec); U_0 for superficial velocity, volumetric flow rate per unit cross-sectional area of packed or fluidized bed (cm/sec). $U_0 = u\varepsilon$

V_b molar volume at the normal boiling point (cm³/g-mol)

V_c catalyst particle volume (cm³)

V_g pore volume per unit mass of catalyst (cm³/g); $V_g = \theta/\rho_p = (1/\rho_p) - (1/\rho_t)$

V_R volume of reactor (cm³)

x distance in direction of diffusion (cm) (Chapter 1). Distance from sealed face (Chapter 3)

x_0 thickness of diffusion layer (cm) (Chapter 1)

X association parameter of solvent (Chapter 1)

Y mole fraction

z distance through packed bed (cm)

Greek

α fraction of total catalyst surface poisoned

β heat generation function $= c_s(-\Delta H)D_{\text{eff}}/\lambda T_s$

γ exponent in Arrhenius reaction rate expression $= E/RT$

$\bar{\Delta}$ average film thickness (Chapter 2)

$\Delta\rho$ difference in density between particle and fluid (g/cm³)

ε void fraction in packed catalyst bed, equals fraction of reactor volume not occupied by catalyst particles

ε Lennard-Jones force constant (Chapter 1)

η effectiveness factor; equals ratio of actual rate of reaction in a porous catalyst to that which would occur if the pellet interior were all exposed to reactants at the same concentration and temperature as that existing at the outside surface of the pellet

θ porosity (void fraction) of catalyst (fraction of gross volume of catalyst pellet that is pore space), $= V_g \rho_p$ (cm³ pores/cm³ pellet)

θ volume change modulus, $= (\nu - 1) Y_{A,s}$ (Section 4.5)

λ thermal conductivity of porous structure or of solid (cal/sec · cm² of total cross section · (°C/cm) = cal/sec · cm · °C)

μ viscosity (poises = g/sec · cm)

ν stoichiometric coefficient

ρ density of fluid; $\Delta\rho$ for difference between density of particle and that of fluid (g/cm³)

ρ_B bulk density of catalyst bed (g/cm³ of reactor volume filled with catalyst particles)

ρ_p particle density (g/cm³ of particle volume); ρ_t for true density of solid material in porous catalyst (g/cm³); ρ_L for density of fluid (g/cm³). $\rho_p = \rho_t(1 - \theta)$

σ Lennard-Jones force constant (Å)

τ tortuosity factor, an empirical factor to correct for "tortuosity" and for nonuniformity of pore cross section; $\tau = D\theta/D_{\text{eff}}$, τ_s for tortuosity factor for surface diffusion; τ_m for tortuosity factor obtained assuming completely Knudsen diffusion and a mean pore radius defined by Equation 1.32 to calculate D_K; τ_p for tortuosity factor calculated by parallel-path pore model

ϕ Thiele diffusion modulus (dimensionless);

$$\phi_s \text{ for sphere} = R\sqrt{\frac{k_v' c_s^{m-1}}{D_{\text{eff}}}} \quad \text{(Equation 3.5).}$$

$$\phi_L \text{ for flat plate geometry} = L\sqrt{\frac{k_v' c_s^{m-1}}{D_{\text{eff}}}} \quad \text{(Equation 3.22)}$$

Φ dimensionless modulus containing only observable or predictable quantities;

$$\Phi_s \text{ for sphere} = \frac{R^2}{D_{\text{eff}}}\left(-\frac{1}{V_c}\frac{dn}{dt}\right)\frac{1}{c_s} \quad \text{(Equation 3.31)}$$

$$\Phi_L \text{ for flat plate geometry} = \frac{L^2}{D_{\text{eff}}}\left(-\frac{1}{V_c}\frac{dn}{dt}\right)\frac{1}{c_s} \quad \text{(Equation 3.32)}$$

ω parameter used for complex or reversible reactions (Equations 4.7 and 4.14)

Mathematical Functions

$\sinh x = \frac{1}{2}(e^x - e^{-x})$

$\cosh x = \frac{1}{2}(e^x + e^{-x})$

$\tanh x = \sinh x/\cosh x$

References

1. Adel'son, S. V., and A. Ya. A. Zaitova, *Khim. i Tekhnol. Topliva Masel*, No. 1 (1962); *Intern. Chem. Eng.*, **2**, 360 (1962); **8** (4), 16 (1963).
2. Adlington, D., and E. Thompson, *Third Symposium on Chemical Reaction Engineering*, Pergamon, New York, 1965, p. 203.
3. Akehata, T., S. Namkoong, H. Kubota, and M. Shindo, *Can. J. Chem. Eng.*, **39**, 127 (1961).
4. Aksel'rud, G. A., *Zh. Fiz. Khim.*, **27**, 10 (1953).
5. Amberg, C. M., and E. Echigoya, *Can. J. Chem. Eng.*, **39**, 215 (1961).
6. Amundson, N. R., and L. R. Raymond, *A.I.Ch.E. J.*, **11**, 339 (1965).
7. Anderson, R. B., F. S. Karn, and J. F. Shultz, *J. Catalysis*, **4**, 56 (1965).
8. Appleby, W. G., J. W. Gibson, and G. M. Good, *Ind. Eng. Chem., Process Design Develop.*, **1**, 102 (1962).
9. Archibald, R. C., N. C. May, and B. S. Greensfelder, *Ind. Eng. Chem.*, **44**, 1811 (1952).
10. Arey, W. G., N. E. Blackwell, III, and A. D. Reichle, *Seventh World Petroleum Congress Proceedings*, Vol. 4, Elsevier, Amsterdam, 1967, p. 167.
11. Aris, R., *Chem. Eng. Sci.*, **6**, 262 (1957).
12. Aris, R., *Ind. Eng. Chem., Fundamentals*, **4**, 227, 487 (1965).
13. Aris, R., *Introduction to the Analysis of Chemical Reactors*, Prentice-Hall, Englewood Cliffs, New Jersey, 1965.
14. Austin, L. G., and P. L. Walker, Jr., *A.I.Ch.E. J.*, **9**, 303 (1963).
15. Barker, J. J., and R. E. Treybal, *A.I.Ch.E. J.*, **6**, 289 (1960).
16. Barnett, L. G., R. E. C. Weaver, and M. M. Gilkeson, *A.I.Ch.E. J.*, **7**, 211 (1961).
17. Baron, T., W. R. Manning, and H. F. Johnstone, *Chem. Eng. Progr.*, **48**, 125 (1952).
18. Barrer, R. M., *Appl. Mater. Res.*, **2** (3), 129 (1963).
19. Barrer, R. M., and J. A. Barrie, *Proc. Roy. Soc.* (*London*), **A213**, 250 (1952).
20. Barrer, R. M., and T. Gabor, *Proc. Roy. Soc.* (*London*), **A251**, 353 (1959).
21. Barrett, E. P., L. G. Joyner, and P. P. Halenda, *J. Am. Chem. Soc.*, **73**, 373 (1951).
22. Basmadjian, D., *J. Catalysis*, **2**, 440 (1963).

23. Battino, R., and H. L. Clever, *Chem. Rev.*, **66**, 395 (1966).
24. Beeck, O., A. E. Smith, and A. Wheeler, *Proc. Roy. Soc.* (*London*), **A117**, 62 (1940).
25. Beek, J., *A.I.Ch.E. J.*, **7**, 337 (1961).
26. Beek, J., *Ind. Eng. Chem., Process Design Develop.*, **1**, 45 (1962).
27. Benesi, H. A., R. V. Bonnar, and C. F. Lee, *Anal. Chem.*, **27**, 1963 (1955).
28. Bieber, H., and E. L. Gaden, paper presented at the Los Angeles meeting of the American Institute of Chemical Engineers, February 7, 1962.
29. Bienert, R., and D. Gelbin, *Chem. Tech.* (*Berlin*), **19**, 207 (1967).
30. Bircumshaw, L. L., and A. C. Riddiford, *Quart. Rev.* (*London*), **6**, 157 (1952).
31. Bird, R. B., "Theory of Diffusion," in *Advances Chem. Eng.*, **1**, 1956, p. 155.
32. Bird, R. B., W. E. Stewart, and E. N. Lightfoot, *Transport Phenomena*, Wiley, New York, 1960.
33. Bischoff, K. B., *A.I.Ch.E. J.*, **11**, 351 (1965).
34. Bischoff, K. B., *Chem. Eng. Sci.*, **22**, 525 (1967).
35. Bischoff, K. B., *Chem. Eng. Sci.*, **23**, 451 (1968).
36. Bischoff, K. B., *Ind. Eng. Chem., Fundamentals*, **5**, 135, 285 (1966).
37. Blue, R. W., V. C. F. Holm, R. B. Regier, E. Fast, and L. Heckelsberg, *Ind. Eng. Chem.*, **44**, 2710 (1952).
38. Bodamer, G., and R. Kunin, *Ind. Eng. Chem.*, **43**, 1082 (1951).
39. Bokhoven, C., and W. van Raayen, *J. Phys. Chem.*, **58**, 471 (1954).
40. Bond, G. C., *Catalysis by Metals*, Academic, New York, 1962, p. 243.
41. Bondi, A., R. S. Miller, and W. G. Schlaffer, *Ind. Eng. Chem., Process Design Develop.*, **1**, 196 (1962).
42. Bowen, J. H., R. Bowrey, and A. S. Malin, *J. Catalysis*, **7**, 209 (1967).
43. Bowman, W. H., Sc.D. thesis, Massachusetts Institute of Technology, 1956.
44. Bradshaw, R. D., and C. O. Bennett, *A.I.Ch.E. J.*, **7**, 48 (1961).
45. Breck, D. W., *J. Chem. Educ.*, **41**, 678 (1964). See also, *Molecular Sieves*, Society of Chemical Industry, London, 1968.
46. Brian, P. L. T., and H. B. Hales, *A.I.Ch.E.J.*, **15**, 419 (1969). Brian, P. L. T., H. B. Hales, and T. K. Sherwood, *A.I.Ch.E. J.*, in press.
47. Brian, P. L. T., K. A. Smith, and L. W. Petri, *Ind. Eng. Chem., Process Design Develop.*, **7**, 21 (1968).
48. Brown, L. F., H. W. Haynes, and W. H. Manogue, *J. Catalysis*, **14**, 220 (1969).
49. Bryant, D. E., and W. L. Kranich, *J. Catalysis*, **8**, 8 (1967).
50. Butt, J. B., *A.I.Ch.E. J.*, **9**, 707 (1963).
51. Butt, J. B., *A.I.Ch.E. J.*, **11**, 106 (1965).
52. Butt, J. B., *Chem. Eng. Sci.*, **21**, 275 (1966).
53. Cadle, P. J., Sc.D. thesis, Massachusetts Institute of Technology, 1966.
54. Cadle, P. J., and C. N. Satterfield, *Ind. Eng. Chem., Fundamentals*, **7**, 189 (1968).
55. Cadle, P. J., and C. N. Satterfield, *Ind. Eng. Chem., Fundamentals*, **7**, 192 (1968).
56. Calderbank, P. H., in *Mixing*, Vol. II, V. W. Uhl and J. B. Gray, Eds., Academic, New York, 1967, Chap. 6.
57. Calderbank, P. H., F. Evans, R. Farley, G. Jepson, and A. Poll, *Proceedings of a Symposium on Catalysis in Practice*, The Institution of Chemical Engineers, London, 1963, p. 66.
58. Calderbank, P. H., and S. J. R. Jones, *Trans. Inst. Chem. Eng.* (*London*), **39**, 363 (1961).
59. Carberry, J. J., *A.I.Ch.E. J.*, **6**, 460 (1960).

60. Carberry, J. J., *A.I.Ch.E. J.*, **7**, 350 (1961).
61. Carberry, J. J., *A.I.Ch.E. J.*, **8**, 557 (1962).
62. Carberry, J. J., *Chem. Eng. Sci.*, **17**, 675 (1962).
63. Carberry, J. J., and R. L. Gorring, *J. Catalysis*, **5**, 529 (1966).
64. Carlton, H. E., and J. H. Oxley, *A.I.Ch.E. J.*, **11**, 79 (1965).
65. Carlton, H. E., and J. H. Oxley, *A.I.Ch.E. J.*, **13**, 87 (1967).
66. Carman, P. C., *Flow of Gases through Porous Media*, Academic, New York, 1956.
67. Chen, N. Y., and P. B. Weisz, *Chem. Eng. Progr. Symp. Ser., No. 73*, **63**, 86 (1967).
68. Chervenak, M. C., E. L. Johanson, C. A. Johnson, and M. Sze, *Oil and Gas J.*, **58** (35), 80, 85 (1960).
69. Chervenak, M. C., C. A. Johnson, and S. C. Schuman, *Petrol. Refiner*, **39** (10), 151 (1960).
70. Chervenak, M. C., C. A. Johnson, and S. C. Schuman, paper presented at the 43rd National Meeting of the American Institute of Chemical Engineers, Tulsa, Oklahoma, September 25–28, 1960.
71. Chu, C., *Ind. Eng. Chem., Fundamentals*, **7**, 509 (1968).
72. Chu, C., and O. A. Hougen, *Chem. Eng. Sci.*, **17**, 167 (1962).
73. Chu, J. C., J. Kalil, and W. A. Wetterath, *Chem. Eng. Progr.*, **49**, 141 (1953).
74. Coenen, J. W. E., in *The Mechanism of Heterogeneous Catalysis*, J. H. DeBoer, Ed., Elsevier, Amsterdam, 1960, p. 126. Also, *Actes du Deuxieme Congres Inter. de Catalyse*, Technip, Paris, 1961, p. 2705.
75. Coenen, J. W. E., H. Boerma, B. G. Linsen, and B. de Vries, *Proceedings of the Third Congress on Catalysis*, W. M. H. Sachtler, G. C. A. Schuit, and P. Zwietering, Eds., North-Holland, Amsterdam, 1965, p. 1387.
76. Corrigan, T. E., J. C. Garver, H. F. Rase, and R. S. Kirk, *Chem. Eng. Progr.*, **49**, 603 (1953).
77. Coughlin, R. W., *A.I.Ch.E. J.*, **13**, 1031 (1967)
78. Cramer, R. H., A. F. Houser, and K. I. Jagel, U. S. Patent 3312615 (April 4, 1967) to Mobil Oil Corp.
79. Crank, J., *The Mathematics of Diffusion*, Oxford, London, 1956.
80. Cranston, R. W., and F. A. Inkley, *Advan. Catalysis*, **9**, 143 (1957).
81. Cunningham, R. A., J. J. Carberry, and J. M. Smith, *A.I.Ch.E. J.*, **11**, 636 (1965).
82. Cunningham, R. E., and J. M. Smith, *A.I.Ch.E. J.*, **9**, 419 (1963).
83. Currie, J. A., *Brit. J. Appl. Phys.*, **11**, 318 (1960).
84. Dacey, J. R., *Ind. Eng. Chem.*, **57** (6), 27 (1965).
85. Damköhler, G., in *Der Chemie-Ingenieur*, Vol. 3, 1937, p. 430.
86. Damköhler, G., *Z. Electrochem.*, **42**, 846 (1936).
87. Damköhler, G., *Z. Physik. Chem.*, **A193**, 16 (1943).
88. Dart, J. C., R. T. Savage, and C. G. Kirkbride, *Chem. Eng. Progr.*, **45**, 102 (1949).
89. Davidson, J. F., and D. Harrison, *Fluidized Particles*, Cambridge, New York, 1963.
90. Davies, W., *Phil. Mag.*, **17**, 235 (1934).
91. Davis, B. R., and D. S. Scott, paper presented at American Institute of Chemical Engineers Meeting, Philadelphia, December 5–9, 1965.
92. Davis, H. S., G. Thomson, and G. S. Crandall, *J. Am. Chem. Soc.*, **54**, 2340 (1932).
93. deAcetis, J., and G. Thodos, *Ind. Eng. Chem.*, **52**, 1003 (1960).
94. Deisler, P. F., Jr., and R. H. Wilhelm, *Ind. Eng. Chem.*, **45**, 1219 (1953).

95. Dixon, J. K., and J. E. Longfield, *Catalysis*, Vol. 7, P. H. Emmett, Ed., Reinhold, New York, 1960, p. 281.
96. Dowden, D. A., and G. W. Bridger, *Advan. Catalysis*, **9**, 669 (1957).
97. Eerkens, J. W., and L. M. Grossman, University of California, Institute of Engineering Research, *Tech. Rept. HE-150–150*, December 5, 1957.
98. Effron, E., and H. E. Hoelscher, *A.I.Ch.E. J.*, **10**, 388 (1964).
99. Essenhigh, R. H., R. Froberg, and J. B. Howard, *Ind. Eng. Chem.*, **57** (9), 32 (1965).
100. Evans, R. B., J. Truitt, and G. M. Watson, *J. Chem. Eng. Data*, **6**, 522 (1961).
101. Evans, R. B., G. M. Watson, and E. A. Mason, *J. Chem. Phys.*, **33**, 2076 (1961).
102. Everett, D. H., and F. S. Stone, Eds., *The Structure and Properties of Porous Materials*, Proceedings of the Tenth Symposium of the Colston Research Society, Academic, New York, 1958.
103. Fan, L-T., Y-C. Yang, and C-Y. Wen, *A.I.Ch.E. J.*, **6**, 482 (1960).
104. Fang, F. T., *Third Congress on Catalysis*, W. M. H. Sachtler, G. C. A. Schuit, and P. Zwietering, Eds., North-Holland, Amsterdam, 1965, p. 901.
105. Field, G. J., H. Watts, and K. R. Weller, *Rev. Pure Appl. Chem.*, **13**, 2 (1963).
106. Foster, R. N., H. Bliss, and J. B. Butt, *Ind. Eng. Chem., Fundamentals*, **5**, 579 (1966).
107. Foster, R. N., and J. B. Butt, *A.I.Ch.E. J.*, **12**, 180 (1966).
108. Foster, R. N., and J. B. Butt, *Ind. Eng. Chem., Fundamentals* **6**, 481 (1967).
109. Foster, R. N., J. B. Butt, and H. Bliss, *J. Catalysis*, **7**, 179, 191 (1967).
110. Frank-Kamenetskii, D. A., *Zh. Tekhn. Fiz.*, **9**, 1457 (1939).
111. Frank-Kamenetskii, D. A., *Diffusion and Heat Exchange in Chemical Kinetics*, 1947, translation into English by N. Thon, Princeton Univ., Princeton, New Jersey, 1955. Second edition, translation into English by J. P. Appleton, Plenum, New York, 1969.
112. Fridland, M. I., *Khim. Teckhnol. Topl. Masel*, No. 5, 7 (1967), *Intern. Chem. Eng.*, **7**, 598 (1967).
113. Friedlander, S. K., *A.I.Ch.E. J.*, **7**, 347 (1961).
114. Frisch, N. W., *Chem. Eng. Sci.*, **17**, 735 (1962).
115. Gamid-Zade, G. A., S. A. Efimova, and N. G. Buzova, *Sb. Tr. Azerb. Nauchn. Issled. Inst. Pererabotke Nefti*, No. 3, 339, 1958; *Chem. Abstr.*, **55**, 27699h (1961).
116. Gardner, L. E., and W. M. Hutchinson, *Ind. Eng. Chem., Prod. Res. Develop.*, **3**, 28 (1964).
117. Gavalas, G. R., *Chem. Eng. Sci.*, **21**, 477 (1966).
118. Gilliland, E. R., R. F. Baddour, and H. H. Engel, *A.I.Ch.E. J.*, **8**, 530 (1962).
119. Gilliland, E. R., R. F. Baddour, and J. L. Russell, *A.I.Ch.E. J.*, **4**, 90 (1958).
120. Gomer, R., R. Wortman, and R. Lundy, *J. Chem. Phys.*, **26**, 1147 (1957).
121. Gorring, R. L., and S. W. Churchill, *Chem. Eng. Progr.*, **57** (7), 53 (1961).
122. Gorring, R. L., and A. J. de Rosset, *J. Catalysis*, **3**, 341 (1964).
123. Graham, R. R., F. C. Vidaurri, Jr., and A. J. Gully, *A.I.Ch.E. J.*, **14**, 473 (1968).
124. Gregg, S. J., and K. S. W. Sing, *Adsorption, Surface Area and Porosity*, Academic, New York, 1967.
125. Grove, A. S., *Ind. Eng. Chem.*, **58** (7), 48 (1966).
126. Gunn, D. J., *Chem. Eng. Sci.*, **21**, 383 (1966).
127. Gunn, D. J., and W. J. Thomas, *Chem. Eng. Sci.*, **20**, 89 (1965).
128. Gupta, V. P., and W. J. M. Douglas, *A.I.Ch.E. J.*, **13**, 883 (1967).

129. Gupta, V. P., and W. J. M. Douglas, *Can. J. Chem. Eng.*, **45**, 117 (1967).
130. Hagerbaumer, W. A., and R. Lee, *Oil and Gas J.*, **45**, 76 (March 15, 1947).
131. Haldeman, R. H., and M. C. Botty, *J. Phys. Chem.*, **63**, 489 (1959).
132. Hall, J. W., and H. F. Rase, *Ind. Eng. Chem., Process Design Develop.*, **2**, 25 (1963).
133. Harriott, P., *A.I.Ch.E. J.*, **8**, 93 (1962).
134. Harriott, P., unpublished paper, 1961.
135. Hartman, J. S., G. W. Roberts, and C. N. Satterfield, *Ind. Eng. Chem., Fundamentals*, **6**, 80 (1967).
136. Hawthorn, R. D., paper presented at American Institute of Chemical Engineers meeting, New York, December 2–7, 1961.
137. Hawtin, P., and R. Murdock, *Chem. Eng. Sci.*, **19**, 819 (1964).
138. Heidel, K., K. Schügerl, F. Fetting, and G. Schiemann, *Chem. Eng. Sci.*, **20**, 557 (1965).
139. Helfferich, F., *Ion Exchange*, McGraw-Hill, New York, 1962.
140. Henry, J. P., B. Chennakesavan, and J. M. Smith, *A.I.Ch.E. J.*, **7**, 10 (1961).
141. Henry, J. P., R. S. Cunningham, and C. J. Geankoplis, *Chem. Eng. Sci.*, **22**, 11 (1967).
142. Hewitt, G. F., and J. R. Morgan, *Progress in Applied Materials Research*, Vol. 5, Gordon and Breach, New York, 1964.
143. Hiemenz, W., *Sixth World Petroleum Congress Proceedings*, Section III, Hanseatische Druckanstalt, Hamburg, 1963, p. 307.
144. Himmelblau, D. M., *Chem. Rev.*, **64**, 527 (1964).
145. Hirsch, C. K., *Molecular Sieves*, Reinhold, New York, 1961.
146. Hirschfelder, J. O., R. B. Bird, and E. L. Spotz, *Chem. Rev.*, **44**, 205 (1949).
147. Hirschfelder, J. O., C. F. Curtiss, and R. B. Bird, *Molecular Theory of Gases and Liquids*, Wiley, New York, 1954.
148. Hixson, A. W., and S. J. Baum, *Ind. Eng. Chem.*, **33**, 478 (1941); **34**, 120 (1942); **36**, 528 (1944).
149. Hlaváček, V., and M. Marek, *Chem. Eng. Sci.*, **23**, 865 (1968).
150. Hlaváček, V., and M. Marek, *Collection Czech. Chem. Commun.*, **32**, 4004 (1967).
151. Hlaváček, V., M. Marek, and M. Kubiček, *Collections Czech. Chem. Commun.*, **33**, 718 (1968).
152. Hobler, T., *Mass Transfer and Absorbers*, Pergamon, London, 1966, English translation.
153. Hoelscher, H. E., *Chem. Eng. Progr. Symp. Ser.*, **50** (10), 45 (1954).
154. Hoog, H., H. G. Klinkert, and A. Schaafsma, *Petrol. Refiner*, **32** (5), 137 (1953).
155. Hoogschagen, J., *Ind. Eng. Chem.*, **47**, 906 (1955).
156. Hughmark, G. A., *Ind. Eng. Chem., Process Design·Develop.*, **6**, 218 (1967).
157. Hutchings, J., and J. J. Carberry, *A.I.Ch.E. J.*, **12**, 20 (1966).
158. Irving, J. P., and J. B. Butt, *Chem. Eng. Sci.*, **22**, 1859 (1967).
159. Johnson, M. F. L., W. E. Kreger, and H. Erickson, *Ind. Eng. Chem.*, **49**, 283 (1957).
160. Johnson, M. F. L., and H. C. Mayland, *Ind. Eng. Chem.*, **47**, 127 (1955).
161. Johnson, M. F. L., and W. E. Stewart, *J. Catalysis*, **4**, 248 (1965).
162. Johnstone, H. F., E. T. Houvouras, and W. R. Schowalter, *Ind. Eng. Chem.*, **46**, 702 (1954).
163. Jost, W., *Diffusion in Solids, Liquids, and Gases*, Academic, New York, 1960.

164. Joyner, L. G., E. P. Barrett, and R. Skold, *J. Am. Chem. Soc.*, **73**, 3158 (1951).
165. Kammermeyer, K., and S.-T. Hwang, *Can. J. Chem. Eng.*, **44**, 82 (1966); *Ind. Eng. Chem., Fundamentals*, **7**, 671 (1968); C. N. Satterfield, *ibid.*, **8**, 175 (1969).
166. Kammermeyer, K. A., and L. O. Rutz, *Chem. Eng. Progr., Symp. Ser.*, **55** (24), 163 (1959).
167. Kao, H. S-P., and C. N. Satterfield, *Ind. Eng. Chem., Fundamentals* **7**, 664 (1968).
168. Katzer, J. R., Sc.D. thesis, Massachusetts Institute of Technology, 1969.
169. Kettenring, K. N., E. L. Manderfield, and J. M. Smith, *Chem. Eng. Progr.*, **46**, 139 (1950).
170. Khoobiar, S., R. E. Peck, and B. J. Reitzer, *Proceedings of the Third International Congress on Catalysis*, W. M. H. Sachtler, G. C. A. Schuit, and P. Zwietering, Eds., North-Holland, Amsterdam, 1965, p. 338. See also R. P. Chambers and M. Boudart, *J. Catalysis*, **6**, 141 (1966).
171. Klimenok, Andreev, and Gordeeva, Academy of Science of the USSR; *Chem. Sci. Bull.*, 521 (1956), in Nestor, J. W., Sc.D. thesis, Massachusetts Institute of Technology, 1964.
172. Knudsen, C. W., G. W. Roberts, and C. N. Satterfield, *Ind. Eng. Chem., Fundamentals*, **5**, 325 (1966).
173. Kolbel, H., and P. Ackermann, *Third World Petroleum Congress Proceedings*, Section N, Paper 2, Brill, Leiden, 1951.
174. Koppers, H., paper presented at the Golden Jubilee Symposium of the Indian Institute of Science, Bangalore, January 11–13, 1960.
175. Kozinski, A. A., and C. J. King, *A.I.Ch.E. J.*, **12**, 109 (1966).
176. Kramer, S. J., *J. Catalysis*, **5**, 190 (1966).
177. Krasuk, J. H., and J. M. Smith, *Ind. Eng. Chem., Fundamentals*, **4**, 102 (1965).
178. Krupiczka, R., *Int. Chem. Eng.*, **7**, 122 (1967); first published in *Chem. Stosowana*, **2B**, 183 (1966).
179. Kubota, H., M. Shindo, T. Akehata, and A. Lin, *Chem. Eng.* (*Japan*), **20**, 11 (1956).
180. Kunii, D., and O. Levenspiel, *Fluidization Engineering*, Wiley, New York, 1969.
181. Kunii, D., and J. M. Smith, *A.I.Ch.E. J.*, **6**, 71 (1960).
182. Kunin, R., *Ion Exchange Resins*, 2nd ed., Wiley, New York, 1958.
183. Kunin, R., E. A. Meitzner, and N. Bortnick, *J. Am. Chem. Soc.*, **84**, 305 (1962).
184. Kunin, R., E. A. Meitzner, J. A. Olive, S. A. Fisher, and N. Frisch, *Ind. Eng. Chem., Process Res. Develop.*, **1**, 140 (1962).
185. Kuo, J. C. W., and N. R. Amundson, *Chem. Eng. Sci.*, **22**, 49, 443, 1185 (1967).
186. Lanneau, K. P., *Trans. Inst. Chem. Eng.*, **38**, 125 (1960).
187. Larson, A. T., and R. S. Tour, *Chem. Met. Eng.*, **26**, 647 (1922).
188. LeNobel, J. W., and J. H. Choufoer, *Fifth World Petroleum Congress Proceedings*, Section III, Paper 18, Fifth World Petroleum Congress, Inc., New York, 1959.
189. Levich, V. G., *Physicochemical Hydrodynamics*, Prentice-Hall, Englewood Cliffs, New Jersey, 1962.
190. Lewis, W. K., E. R. Gilliland, and W. Glass, *A.I.Ch.E. J.*, **5**, 419 (1959).
191. Lochiel, A. C., and P. H. Calderbank, *Chem. Eng. Sci.*, **19**, 471 (1964).
192. Love, K. S., and P. H. Emmett, *J. Am. Chem. Soc.*, **63**, 3297 (1941).
193. Luss, D., *Can. J. Chem. Eng.*, **46**, 154 (1968).
194. Luss, D., and N. R. Amundson, *A.I.Ch.E. J.*, **13**, 759 (1967).

195. Luss, D., and N. R. Amundson, *A.I.Ch.E. J.*, **14**, 211 (1968).
196. Luss, D., and N. R. Amundson, *Chem. Eng. Sci.* **22**, 253 (1967).
197. Luss, D., and N. R. Amundson, *Ind. Eng. Chem., Fundamentals*, **6**, 457 (1967).
198. Marek, M., and V. Hlaváček, *Collection Czech. Chem. Commun.*, **33**, 506 (1968).
199. Mars, P., and M. J. Gorgels, *Proceedings of the Third European Symposium on Chemical Reaction Engineering*, Pergamon, New York, 1965, p. 55.
200. Masamune, S., and J. M. Smith, *A.I.Ch.E. J.*, **8**, 217 (1962).
201. Masamune, S., and J. M. Smith, *A.I.Ch.E. J.*, **12**, 384 (1966).
202. Masamune, S., and J. M. Smith, *Ind. Eng. Chem., Fundamentals*, **2**, 137 (1963).
203. Masamune, S., and J. M. Smith, *J. Chem. Eng. Data*, **8**, 54 (1963).
204. Massoth, F. E., *Ind. Eng. Chem., Process Design Develop.*, **6**, 200 (1967).
205. Mathis, J. F., and C. C. Watson, *A.I.Ch.E. J.*, **2**, 518 (1959).
206. Mattern, R. V., O. Bilous, and E. L. Piret, *A.I.Ch.E. J.*, **3**, 497 (1957).
207. Maxted, E. B., and J. S. Elkins, *J. Chem. Soc.*, 5086 (1961).
208. Maxwell, J. B., *Data Book on Hydrocarbons*, Van Nostrand, Princeton, New Jersey, 1950.
209. May, W. G., *Chem. Eng. Progr.*, **55** (12), 49 (1959).
210. Maymo, J. A., R. E. Cunningham, and J. M. Smith, *Ind. Eng. Chem., Fundamentals*, **5**, 280 (1966).
211. Maymo, J. A., and J. M. Smith, *A.I.Ch.E. J.*, **12**, 845 (1966).
212. McAdams, W. H., *Heat Transmission*, 3rd ed., McGraw-Hill, New York, 1954.
213. McGuire, M. L., and L. Lapidus, *A.I.Ch.E. J.*, **11**, 85 (1965).
214. Miale, J. N., N. Y. Chen, and P. B. Weisz, *J. Catalysis*, **6**, 278 (1966).
215. Michaels, A. S., *A.I.Ch.E. J.*, **5**, 270 (1959).
216. Mickley, H. S., J. W. Nestor, and L. A. Gould, *Can. J. Chem. Eng.*, **43**, 61 (1965).
217. Miller, D. N., and R. S. Kirk, *A.I.Ch.E. J.*, **8**, 183 (1962).
218. Miller, F. W., and H. A. Deans, *A.I.Ch.E. J.*, **13**, 45 (1967).
219. Mingle, J. O., and J. M. Smith, *A.I.Ch.E. J.*, **7**, 243 (1961).
220. Mischke, R. A., and J. M. Smith, *Ind. Eng. Chem., Fundamentals*, **1**, 288 (1962).
221. Mitschka, P., and P. Schneider, submitted for publication.
222. Mitschka, P., P. Schneider, and V. S. Beskov, *Collection Czech. Chem. Commun.*, **33**, 3598 (1968).
223. Murakami, Y., T. Kobayashi, T. Hattori, and M. Masuda, *Ind. Eng. Chem., Fundamentals*, **7**, 599 (1968).
224. Nagata, S., I. Yamaguchi, S. Yabuta, and M. Harada, *Mem. Fac. Eng. Kyoto Univ.*, **22**, 86 (1960).
225. Newham, J., and R. L. Burwell, Jr., *J. Phys. Chem.*, **66**, 1431, 1438 (1962).
226. Nichols, J. R., Ph.D. thesis, The Pennsylvania State University, 1961.
227. Nielsen, A., *An Investigation on Promoted Iron Catalysts for the Synthesis of Ammonia*, 3rd ed., Gjellerups, Copenhagen, 1968.
228. Norton, C. J., *Ind. Eng. Chem., Prod. Res. Develop.*, **3**, 230 (1964).
229. O'Connell, F. P., and D. E. Mack, *Chem. Eng. Progr.*, **46**, 358 (1950).
230. O'Connell, J. E., Sc.D. thesis, Massachusetts Institute of Technology, 1964.
231. Oele, A. P., *First European Symposium on Chemical Engineering*, Pergamon, New York, 1957, p. 146.
232. Olander, D. R., *Ind. Eng. Chem., Fundamentals*, **6**, 178, 188 (1967).
233. Olson, J. H., *Ind. Eng. Chem., Fundamentals*, **7**, 185 (1968).

234. Olson, K. E., D. Luss, and N. R. Amundson, *Ind. Eng. Chem., Process Design Develop.*, **7**, 96 (1968).
235. Orcutt, J. C., J. F. Davidson, and R. L. Pigford, *Chem. Eng. Progr., Symp. Ser., No. 38*, **58**, 1 (1962).
236. Osberg, G. L., A. Tweddle, and W. C. Brennan, *Can. J. Chem. Eng.*, **41**, 260 (1963).
237. Ostergaard, K., *Acta Chem. Scand.*, **15**, 2037 (1961) (in English).
238. Ostergaard, K., *Chem. Eng. Sci.*, **18**, 259 (1963).
239. Ostergaard, K., *Proceedings of the Third Congress on Catalysis*, W. M. H. Sachtler, G. C. A. Schuit, and P. Zwietering, Eds., Wiley, New York, 1965, p. 1348.
240. Otani, S., and J. M. Smith, *J. Catalysis*, **5**, 332 (1966).
241. Otani, S., N. Wakao, and J. M. Smith, *A.I.Ch.E. J.*, **10**, 130 (1964).
242. Otani, S., N. Wakao, and J. M. Smith, *A.I.Ch.E. J.*, **11**, 439 (1965).
243. Otani, S., N. Wakao, and J. M. Smith, *A.I.Ch.E. J.*, **11**, 446 (1965).
244. Ozawa, Y., and K. B. Bischoff, *Ind. Eng. Chem., Process Design Develop.*, **7**, 67 (1968).
245. Pansing, W. F., *A.I.Ch.E. J.*, **2**, 71 (1956).
246. Pasquon, I., and M. Dente, *J. Catalysis*, **1**, 508 (1962).
247. Peters, C., and R. Krabetz, *Z. Elektrochem.*, **60**, 859 (1956).
248. Petersen, E. E., *A.I.Ch.E. J.*, **4**, 343 (1958).
249. Petersen, E. E., *Chem. Eng. Sci.*, **17**, 987 (1962).
250. Petersen, E. E., *Chem. Eng. Sci.*, **20**, 587 (1965).
251. Petersen, E. E., *Chemical Reaction Analysis*, Prentice-Hall, New York, 1965.
252. Petrovic, L. J., and G. Thodos, *Ind. Eng. Chem., Fundamentals*, **7**, 274 (1968).
253. Pohlhausen, E., *Z. Angew. Math. Mech.*, **1**, 115 (1921).
254. Polejes, J. D., Ph.D. thesis, University of Wisconsin, 1959; see also Johnson, D. L., H. Saito, J. D. Polejes, and O. A. Hougen, *A.I.Ch.E. J.*, **3**, 411 (1957).
255. Pollard, W. G., and R. D. Present, *Phys. Rev.*, **73**, 762 (1948).
256. Prater, C. D., *Chem. Eng. Sci.*, **8**, 284 (1958).
257. Prater, C. D., and R. M. Lago, *Advan. Catalysis*, **VIII**, 293 (1956).
258. Ranz, W. E., and W. R. Marshall, *Chem. Eng. Progr.*, **48**, 141 (1952).
259. Rao, M. R., and J. M. Smith, *A.I.Ch.E. J.*, **9**, 485 (1963).
260. Rao, M. R., and J. M. Smith, *A.I.Ch.E. J.*, **10**, 293 (1964).
261. Rao, M. R., N. Wakao, and J. M. Smith, *Ind. Eng. Chem., Fundamentals*, **3**, 127 (1964).
262. Reamer, H. H., and B. H. Sage, *A.I.Ch.E. J.*, **3**, 449 (1957).
263. Reed, E. W., and J. S. Dranoff, *Ind. Eng. Chem., Fundamentals*, **3**, 304 (1964).
264. Reid, R. C., and T. K. Sherwood, *Properties of Gases and Liquids*, 2nd ed., McGraw-Hill, New York, 1966.
265. Reif, A. E., *J. Phys. Chem.*, **56**, 778 (1952).
266. Riccetti, R. E., and G. Thodos, *A.I.Ch.E. J.*, **7**, 442 (1961).
267. Richardson, J. F., and P. Ayers, *Trans. Inst. Chem. Eng.*, **37**, 314 (1959).
268. Richardson, J. F., and A. G. Bakhtiar, *Trans. Inst. Chem. Eng.*, **36**, 283 (1958).
269. Richardson, J. F., and J. Szekeley, *Trans. Inst. Chem. Eng.*, **39**, 212 (1961).
270. Ritter, H. L., and L. E. Drake, *Ind. Eng. Chem., Anal. Ed.*, **17**, 782, 787 (1945).
271. Riverola, J. B., and J. M. Smith, *Ind. Eng. Chem., Fundamentals*, **3**, 308 (1964).
272. Roberts, G. W., and C. N. Satterfield, *Ind. Eng. Chem., Fundamentals*, **4**, 288 (1965).

273. Roberts, G. W., and C. N. Satterfield, *Ind. Eng. Chem., Fundamentals*, **5**, 317 (1966).
274. Robertson, J. L., and J. M. Smith, *A.I.Ch.E. J.*, **9**, 342 (1963).
275. Rockett, T. J., W. R. Foster, and R. G. Ferguson, Jr., *J. Am. Ceram. Soc.*, **48**, 329 (1965).
276. Rosner, D. E., *Chem. Eng. Sci.*, **19**, 1 (1964); *A.I.Ch.E. J.*, **9**, 289 (1963).
277. Ross, L. D., *Chem. Eng. Progr.*, **61** (10), 77 (1965).
278. Rothfeld, L. B., *A.I.Ch.E. J.*, **9**, 19 (1963).
279. Rowe, P. N., *Chem. Eng. Progr.*, **60** (3), 75 (1964).
280. Rowe, P. N., *Chem. Eng. Progr., Symp. Ser., No. 38*, **58**, 42 (1962).
281. Rowe, P. N., and K. T. Claxton, Reports AERE-R4673 and AERE-R4675, Chemical Engineering Division, Atomic Energy Establishment, Harwell, England, 1964.
282. Rowlinson, J. S., and J. R. Townley, *Trans. Faraday Soc.*, **94**, 20 (1953).
283. Rozovskii, A. Ya., and V. V. Shchekin, *Kinetika i Kataliz*, **1**, 313 (1960) [English transl.: *Kinetics Catalysis* (*USSR*), **1**, 286 (1960)].
284. Rozovskii, A. Ya., V. V. Shchekin, and E. G. Pokrovskaya, *Kinetika i Kataliz*, **1**, 464 (1960) [English transl.: *Kinetics Catalysis* (*USSR*), **1**, 432 (1960)].
285. Rudershausen, C. G., and C. C. Watson, *Chem. Eng. Sci.*, **3**, 110 (1954).
286. Rushton, J. H., E. W. Costich, and H. J. Everett, *Chem. Eng. Progr.*, **46**, 395, 467 (1950).
287. Sada, E., and C-Y. Wen, *Chem. Eng. Sci.*, **22**, 559 (1967).
288. Sagara, M., S. Masamune, and J. M. Smith, *A.I.Ch.E. J.*, **13**, 1226 (1967).
289. Saito, H., F. Shimamoto, Y. Mishima, and O. Sataka, *Kogyo Kagaku Zasshi*, **64**, 1733, 1738 (1961).
290. Saraf, S. K., Sc.D. thesis, Massachusetts Institute of Technology, 1964.
291. Satterfield, C. N., and F. P. Audibert, *Ind. Eng. Chem., Fundamentals*, **2**, 200 (1963).
292. Satterfield, C. N., and P. J. Cadle, *Ind. Eng. Chem., Fundamentals*, **7**, 202 (1968).
293. Satterfield, C. N., and P. J. Cadle, *Ind. Eng. Chem., Process Design Devel.*, **7**, 256 (1968).
294. Satterfield, C. N., and F. Feakes, *A.I.Ch.E. J.*, **5**, 122 (1959).
295. Satterfield, C. N., and A. J. Frabetti, Jr., *A.I.Ch.E. J.*, **13**, 731 (1967).
296. Satterfield, C. N., and H. Iino, *Ind. Eng. Chem., Fundamentals*, **7**, 214 (1968).
297. Satterfield, C. N., Y. H. Ma, and T. K. Sherwood, *Trans. Inst. Chem. Eng.*, in press.
298. Satterfield, C. N., A. A. Pelossof, and T. K. Sherwood, *A.I.Ch.E. J.*, **15**, 226 (1969).
299. Satterfield, C. N., and H. Resnick, *Chem. Eng. Progr.*, **50**, 504 (1954).
300. Satterfield, C. N., and S. K. Saraf, *Ind. Eng. Chem., Fundamentals*, **4**, 451 (1965).
301. Satterfield, C. N., and R. S. C. Yeung, *Ind. Eng. Chem., Fundamentals*, **2**, 257 (1963).
302. Schilson, R. E., and N. R. Amundson, *Chem. Eng. Sci.*, **13**, 226, 237 (1961).
303. Schneider, P., and P. Mitschka, *Chem. Eng. Sci.*, **21**, 455 (1966).
304. Schneider, P., and P. Mitschka, *Collection Czech. Chem. Commun.*, **30**, 146 (1965), in English.
305. Schneider, P., and P. Mitschka, *Collection Czech. Chem. Commun.*, **31**, 1205 (1966), in English.

306. Schneider, P., and P. Mitschka, *Collection Czech. Chem. Commun.*, **31**, 3677 (1966), in English.
307. Schneider, P., and P. Mitschka, "Fourth International Congress on Catalysis," Moscow, 1968, preprint.
308. Schneider, P., and J. M. Smith, *A.I.Ch.E. J.*, **14**, 886 (1968).
309. Schuman, S. C., *Chem. Eng. Progr.*, **57** (12), 49 (1961).
310. Schwertz, F. A., *J. Am. Ceram. Soc.*, **32**, 390 (1949).
311. Scott, D. S., and K. E. Cox, *Can. J., Chem. Eng.*, **3**, 201 (1960).
312. Scott, D. S., and F. A. L. Dullien, *A.I.Ch.E. J.*, **8**, 113 (1962).
313. Sehr, R. A., *Chem. Eng. Sci.*, **9**, 145 (1958).
314. Sen Gupta, A., and G. Thodos, *A.I.Ch.E. J.*, **8**, 608 (1962).
315. Sen Gupta, A., and G. Thodos, *Chem. Eng. Progr.*, **58** (7), 58 (1962).
316. Shen, C. Y., and H. F. Johnstone, *A.I.Ch.E. J.*, **1**, 349 (1955).
317. Sherwood, T. K., and E. J. Farkas, *Chem. Eng. Sci.*, **21**, 573 (1966).
318. Sherwood, T. K., and F. A. Holloway, *Trans. Am. Inst. Chem. Engrs.*, **36**, 21 (1940).
319. Sherwood, T. K., and R. L. Pigford, *Absorption and Extraction*, McGraw-Hill, New York, 1952.
320. Shingu, H., U. S. Patent 2985668 (May 23, 1961).
321. Shulman, H. L., C. F. Ullrich, and N. Wells, *A.I.Ch.E. J.*, **1**, 247 (1955).
322. Sinfelt, J. H., *Advan. Chem. Eng.*, **5**, 37 (1964).
323. Sladek, K. J., Sc.D. thesis, Massachusetts Institute of Technology, 1967.
324. Slattery, J. C., and R. B. Bird, *A.I.Ch.E. J.*, **4**, 137 (1958).
325. Smith, N. L., and N. R. Amundsen, *Ind. Eng. Chem.*, **43**, 2156 (1951).
326. Sokol'skii, D. V., *Hydrogenation in Solutions*, English translation by N. Kaner and J. A. Epstein, Davey, New York, 1964, pp. 365–371.
327. Steinberger, R. L., and R. E. Treybal, *A.I.Ch.E. J.*, **6**, 227 (1960).
328. Steisel, N., and J. B. Butt, *Chem. Eng. Sci.*, **22**, 469 (1967).
329. Sterrett, J. S., and L. F. Brown, *A.I.Ch.E. J.*, **14**, 696 (1968).
330. Suga, K., Y. Morita, E. Kunugita, and T. Otake, *Kogyo Kagaku Zasshi*, No. 2, 136 (1967); *Int. Chem. Eng.*, **7**, 742 (1967).
331. Svehla, R. A., Tech. Rept. R-132, Lewis Research Center, NASA, Cleveland, 1962.
332. Thiele, E. W., *Am. Scientist*, **55** (2), 176 (1967).
333. Thiele, E. W., *Ind. Eng. Chem.*, **31**, 916 (1939).
334. Thomas, J. M., and W. J. Thomas, *Introduction to the Principles of Heterogeneous Catalysis*, Academic, New York, 1967, pp. 338–363.
335. Tinkler, J. D., and A. B. Metzner, *Ind. Eng. Chem.*, **53**, 663 (1961); see also D. Luss, *A.I.Ch.E. J.*, **14**, 966 (1968).
336. Tinkler, J. D., and R. L. Pigford, *Chem. Eng. Sci.*, **15**, 326 (1961).
337. Topchieva, K. V., T. V. Antipina, and L. H. Shien, *Kinetika i Kataliz*, **1**, 471 (1960) [English transl.: *Kinetics Catalysis*, **1**, 438 (1960)]. Also, *Actes du Deuxieme Congres Inter. de Catalyse*, Technip, Paris, 1961, p. 1955.
338. Tschernitz, J., S. Bornstein, R. Beckmann, and O. A. Hougen, *Trans. A.I.Ch.E.*, **42**, 883 (1946).
339. Tu, C. M., H. Davis, and H. C. Hottel, *Ind. Eng. Chem.* **26**, 749 (1934).
340. Vaidyanathan, K., and L. K. Doraiswamy, *Chem. Eng. Sci.*, **23**, 537 (1968).
341. Valentin, F. H. H., *Absorption in Gas-Liquid Dispersions*, E. & F. N. Spon, London, 1967.

342. Van Deemter, J. J., *Chem Eng. Sci.*, **13**, 143 (1961).
343. Van Deemter, J. J., *Third Symposium on Chemical Reaction Engineering*, Pergamon, New York, 1965, p. 215.
344. Van Krevelen, D. W., and J. T. C. Krekels, *Rec. Trav. Chim.*, **67**, 512 (1948).
345. Van Zoonen, D., *Third Congress on Catalysis*, W. M. H. Sachtler, G. C. A. Schuit, and P. Zwietering, Eds., North-Holland, Amsterdam, 1965, p. 1319.
346. Van Zoonen, D., and C. Th. Douwes, *J. Inst. Petrol.*, **49** (480), 383 (1963).
347. Venuto, P. B., and L. A. Hamilton, *Ind. Eng. Chem., Prod. Res. Develop.*, **6**, 190 (1967).
348. Villet, R. H., and R. H. Wilhelm, *Ind. Eng. Chem.*, **53**, 837 (1961).
349. Vinograd, J. R., and J. W. McBain, *J. Am. Chem. Soc.*, **63**, 2008 (1941).
350. Vlasenko, V. M., M. T. Rusov, and G. E. Yuzefovich, *Kinetika i Kataliz*, **2**, 525 (1961) [English transl.: *Kinetics Catalysis*, **2**, 476 (1961)].
351. Wagner, C., *Chem. Tech.* (*Berlin*), **18**, 1, 28 (1945).
352. Wagner, C., *Z. Physik. Chem.*, **A193**, 1 (1943).
353. Wakao, N., S. Otani, and J. M. Smith, *A.I.Ch.E. J.*, **11**, 435 (1965).
354. Wakao, N., P. W. Selwood, and J. M. Smith, *A.I.Ch.E. J.*, **8**, 478 (1962).
355. Wakao, N., and J. M. Smith, *Chem. Eng. Sci.*, **17**, 825 (1962).
356. Wakao, N., and J. M. Smith, *Ind. Eng. Chem., Fundamentals*, **3**, 123 (1964).
357. Walker, P. L., Jr., L. G. Austin, and S. P. Nandi, in *Chemistry and Physics of Carbon*, Vol. 2, P. L. Walker, Jr., Ed., Dekker, New York, 1966, p. 257.
358. Walker, P. L., Jr., F. Rusinko, Jr., and L. G. Austin, *Advan. Catalysis*, **XI**, 133 (1959).
359. Ware, C. H., Jr., Ph.D. thesis, University of Pennsylvania, 1959.
360. Weaver, J. A., and A. B. Metzner, *A.I.Ch.E. J.*, **12**, 655 (1966).
361. Wechsler, A., Sc.D. thesis, Massachusetts Institute of Technology, 1961; Satterfield, C. N., R. C. Reid, and A. E. Wechsler, *A.I.Ch.E. J.*, **9**, 168 (1963).
362. Weekman, V. W., Jr., *J. Catalysis*, **5**, 44 (1966).
363. Weekman, V. W., Jr., and R. L. Gorring, *J. Catalysis*, **4**, 260 (1965).
364. Wei, J., *Chem. Eng. Sci.*, **20**, 729 (1965).
365. Wei, J., *Chem. Eng. Sci.*, **21**, 1171 (1966).
366. Wei, J., *J. Catalysis*, **1**, 526, 538 (1962).
367. Wei, J., and C. D. Prater, *Advan. Catalysis*, **XIII**, 233 (1962).
368. Weisz, P. B., *Advan. Catalysis*, **XIII**, 137 (1962).
369. Weisz, P. B., *Chem. Eng. Progr. Symp. Ser.*, **55** (25), 29 (1959).
370. Weisz, P. B., *J. Catalysis*, **6**, 425 (1966).
371. Weisz, P. B., *Science*, **123**, 887 (1956).
372. Weisz. P. B., *Z. Physik Chem., Neue Folge*, **11**, 1 (1957), in English.
373. Weisz, P. B., V. J. Frilette, R. W. Maatman, and E. B. Mower, *J. Catalysis*, **1**, 307 (1962).
374. Weisz, P. B., and R. D. Goodwin, *J. Catalysis*, **2**, 397 (1963).
375. Weisz, P. B., and R. D. Goodwin, *J. Catalysis*, **6**, 227 (1966).
376. Weisz, P. B., and J. S. Hicks, *Chem. Eng. Sci.*, **17**, 265 (1962).
377. Weisz, P. B., and C. D. Prater, *Advan. Catalysis*, **VI**, 143 (1954).
378. Weisz, P. B., and A. B. Schwartz, *J. Catalysis*, **1**, 399 (1962).
379. Weisz, P. B., and E. W. Swegler, *J. Phys. Chem.*, **59**, 823 (1955).
380. Weisz, P. B., and E. W. Swegler, *Science*, **126**, 31 (1957).
381. Whang, H. Y., Sc.D. thesis, Massachusetts Institute of Technology, 1961.
382. Wheeler, A., in *Advan. Catalysis*, **III**, 249 (1951).

383. Wheeler. A., in *Catalysis*, Vol. 2, P. H. Emmett, Ed., Reinhold, New York, 1955.
384. Wicke, E., *Chem.-Ing.-Tech.*, **29**, 305 (1957).
385. Wicke, E., and W. Brötz, *Chem.-Ing.-Tech.*, **21**, 219 (1949).
386. Wicke, E., and K. Hedden, *Z. Elektrochem.*, **57**, 636 (1953).
387. Wicke, E., and R. Kallenbach, *Kolloid. Z.*, **97**, 135 (1941).
388. Wiggs, P. K. C., in *The Structure and Properties of Porous Materials*, D. H. Everett and F. S. Stone, Eds., Academic, New York, 1958, p. 183.
389. Wilhelm, R. H., *Pure Appl. Chem.*, **5**, 403 (1962).
390. Wilke, C. R., and P. Chang, *A.I.Ch.E. J.*, **1**, 264 (1955).
391. Wilson, E. J., and C. J. Geankoplis, *Ind. Eng. Chem., Fundamentals*, **5**, 9 (1966).
392. Winegardner, D. K., and R. A. Schmitz, *A.I.Ch.E. J.*, **14**, 301 (1968).
393. Wu, P. C., Sc.D. thesis, Massachusetts Institute of Technology, 1949.
394. Yeh, G. C., *J. Chem. Eng. Data*, **6**, 526 (1961).
395. Yoshida, F., D. Ramaswami, and O. A. Hougen, *A.I.Ch.E. J.*, **8**, 5 (1962).
396. Zajcew, M., *Am. Oil Chemists' Soc. J.* **37**, 11 (1960).
397. Zeldovitch, Ya. B., *Acta Physicochim. U.R.S.S.*, **10**, 583 (1939), in English. Also, *Z. Fiz. Chim.*, **13**, 163 (1939).
398. Zhorov, Y. M., G. M. Panchenkov, M. E. Levinter, M. E. Kozlov, and B. F. Morozov, *Russian J. Phys. Chem.*, **40**, 290 (1966), English translation.

Subject Index

Author Index

To assist in locating information, when a reference is cited in the text without an author's name, the reference number is shown here in parentheses following the number of the page on which it appears.